T0400075

A Step Towards Society 5.0

Demystifying Technologies for Computational Excellence: Moving Towards Society 5.0

Series Editors: Vikram Bali and Vishal Bhatnagar

This series encompasses research work in the field of Data Science, Edge Computing, Deep Learning, Distributed Ledger Technology, Extended Reality, Quantum Computing, Artificial Intelligence, and various other related areas, such as natural-language processing and technologies, high-level computer vision, cognitive robotics, automated reasoning, multivalent systems, symbolic learning theories and practice, knowledge representation and the semantic web, intelligent tutoring systems, AI and education.

The prime reason for developing and growing this new book series is to focus on the latest technological advancements: their impact on the society, the challenges faced in implementation, and the drawbacks or reverse impact on the society due to technological innovations. With the technological advancements, every individual has personalized access to all the services, with all devices connected with each other communicating amongst themselves, thanks to this technology for making our lives simpler and easier. These aspects will help us to overcome the drawbacks of the existing systems and help in building new systems with the latest technologies that will help the society in various ways, proving Society 5.0 to be one of the biggest revolutions of this era.

Artificial Intelligence, Machine Learning, and Data Science Technologies
Future Impact and Well-Being for Society 5.0
Edited by Neeraj Mohan, Ruchi Singla, Priyanka Kaushal, and Seifedine Kadry

Transforming Higher Education Through Digitalization
Insights, Tools, and Techniques
Edited by S. L. Gupta, Nawal Kishor, Niraj Mishra, Sonali Mathur, and Utkarsh Gupta

A Step Towards Society 5.0
Research, Innovations, and Developments in Cloud-Based Computing Technologies
Edited by Shahnawaz Khan, Thirunavukkarasu K., Ayman AlDmour, and Salam Salameh Shreem

For more information on this series, please visit: www.routledge.com/Demystifying-Technologies-for-Computational-Excellence-Moving-Towards-Society-5.0/book-series/CRCDTCEMTS

A Step Towards Society 5.0

Research, Innovations, and Developments in Cloud-Based Computing Technologies

Edited by
Shahnawaz Khan,
Thirunavukkarasu K.,
Ayman AlDmour, and
Salam Salameh Shreem

CRC Press is an imprint of the
Taylor & Francis Group, an **informa** business

First edition published 2022
by CRC Press
6000 Broken Sound Parkway NW, Suite 300, Boca Raton, FL 33487-2742

and by CRC Press
2 Park Square, Milton Park, Abingdon, Oxon OX14 4RN

CRC Press is an imprint of Taylor & Francis Group, LLC

Library of Congress Cataloging-in-Publication Data
Names: Khan, Shahnawaz, editor.
Title: A step towards society 5.0: research, innovations, and developments in
cloud-based computing technologies/edited by Shahnawaz Khan,
Thirunavukkarasu K., Ayman AlDmour, and Salam Salameh Shreem.
Description: Boca Raton: CRC Press, 2022. |
Series: Demystifying technologies for computational excellence |
Includes bibliographical references and index.
Identifiers: LCCN 2021020178 (print) | LCCN 2021020179 (ebook) |
ISBN 9780367685461 (hbk) | ISBN 9780367685485 (pbk) | ISBN 9781003138037 (ebk)
Subjects: LCSH: Cloud computing. | Industry 4.0.
Classification: LCC QA76.585 .S847 2022 (print) |
LCC QA76.585 (ebook) | DDC 004.67/82–dc23
LC record available at https://lccn.loc.gov/2021020178
LC ebook record available at https://lccn.loc.gov/2021020179

ISBN: 978-0-367-68546-1 (hbk)
ISBN: 978-0-367-68548-5 (pbk)
ISBN: 978-1-003-13803-7 (ebk)

DOI: 10.1201/9781003138037

Typeset in Times
by Newgen Publishing UK

Contents

Preface

Cloud technology has achieved popularity and drawn interest from the worldwide community. For business organizations, it is sometimes difficult to manage the demands of customers and keep pace with the technology. Therefore, there are several complementary and converging factors that are giving a boost to the rise of cloud technology. Technology giants like Amazon, Google, and Microsoft have been innovating and developing cloud technology and its applications for almost two decades now. However, it has received recognition mainly in the last five years. The maturity of cloud technology and cloud services has been increasing every day. Similarly, awareness about the benefits and limitations of cloud technology is also supporting its adoption and the migration to cloud technology. It can help businesses to scale up and down in a very short period of time. It benefits the businesses to think about scaling without thinking much about the information technology requirements. Cloud technology is also bridging the gap (digital divide) between small organizations that cannot boast their IT infrastructures and big business, by removing the infeasibility of the deployments of the new applications.

The motive of this book is to give insightful information on the cutting-edge innovations in cloud technology. It contains a wealth of information for developing cloud technology knowledge in various domains of application. The book focuses on developing an understanding of the current and future innovations in cloud technology and explores its applications and pioneering innovations. It will help the researchers and learners to engage in deep and meaningful conversations with their peers, working in cloud technology. It will also support the development of an understanding of cutting-edge innovations, paradigms, and security, by using real-life applications, case studies, and examples.

This book covers not only theoretical approaches, algorithms but contains a sequence of steps to the analysis of problems with data, process, reports, and optimization techniques. It comprises of a variety of real-life applications, based on emerging trends of cloud technology and machine learning. It also explores certain aspects of cloud technology from the research, scientific, and business perspective, for secure and scalable applications in various fields such as in Society 5.0, next-generation computing excellence, and so on.

Preface

Editor Biographies

Shahnawaz Khan is working as the Secretary-General of the Scientific Research Council at the University College of Bahrain. He holds a Ph.D. (Computer Science) from the Indian Institute of Technology (BHU), India. He owns 3 patents and has published 34 research papers in refereed journals and international conferences. His research interests are in artificial intelligence, natural language processing, blockchain, and cloud computing. He has been guest editor for Scopus and SCI-indexed journals and has edited several books. He has participated in many international conferences as a keynote speaker and member of the International Program/technical Committees. He is responsible for preparing, budgeting, and executing institutional research plans and policy and is working with the Ministry of Higher Education for institutional research growth, and innovative technology transfer for the university. His activities also comprise academic, research leadership, administration, faculty development, budget preparation & management, curricula, syllabi, laboratory and infrastructure development, collaboration building with knowledge institutions and with industry. He is actively involved in program accreditations (ABET, BQA, NCAAA, and NACC).

Thirunavukkarasu K. is a distinguished academician with over twenty-three years of experience in teaching and in the software industry. At the moment, he is heading the Department of Data Science in Karnavati University. He has a Bachelor's Degree in Computer Science from the University of Madras, and 3 Master's degrees in Computer Science and Engineering, a Ph.D. and has completed post-doctoral work . He has filed 5 patents in IPR, Govt. of India. He has presented/published 24 research papers at both National and International level, which includes contributions to Scopus, IEEE Xplore and Springer. He has various appraisals and achievements to his credit which includes a best faculty award, IBM Coordinator to float specialized programs. He has certifications to his credit from Oracle, IBM, and VMware in the courses of BI, DB2, Data Mining and Predictive modeling, and Cloud Computing. He also has various professional certifications to his credit and is an active member of various professional societies which include IEEE, IAENG, LMCSI, and LMISTE. He has actively involved in academic activities, lab setup in the campus and course alignment with industry, OBE process, NBA and ABET accreditation.

Ayman AlDmour is working as Acting Dean of the College of Arts and Science at the Applied Science University. He has obtained his BSc in Electronics – Communication Engineering in 1993 from the Jordan University of Science and Technology. He has worked as a Telecom Engineer in Orange, Jordan from 1996 to 2006. He has pursued his MSc and PhD in 2003 and 2006, respectively, in Computer Information Systems at the Arab Academy of Banking and Financial Sciences, Jordan. In Feb 2006, he joined the College of Information Technology at Al-Hussein Bin Talal University (AHU), Jordan. At AHU, he chaired the Department of Computer Information Systems, the Computer and Information Technology Centre, and the College of Information

Technology. In March 2016, he was appointed a President Consultant for Scientific Affairs (acting as Vice-president for Academic Affairs) and President's Assistant for Central Bids. He joined the Applied Science University as Acting Dean of the College of Arts and Science in September 2018.

Salam Salameh Shreem received his Ph.D in Computer Science at Universiti Kebangsaan Malaysia (UKM) in January 2014 and his BSc degree in Computer Information Systems from the Al-Zaytoonah University of Jordan, Amman, Jordan, in 2004, and his MSc degree in Computer Science from Al-Balqa' Applied University, Al-Salt, Jordan, in 2007. His research interest involves data mining and metaheuristic algorithms for combinatorial optimisation problems such as the gene selection problem, university timetabling and job shop scheduling. He is currently working as an IT consultant with HLT Service Group in Illinois, United States.

1 Digitalization to Society 5.0

The New Paradigm

Mohammad Haider Syed

CONTENTS

1.1 INTRODUCTION

Human evolution and its needs have forced humans to create innovative ideas and artefacts. With the development of the scriptures, people started recording their data on stones, and the like. These scriptures are of immense value in understanding the evolution of humankind. These necessities stretched humans to their limits and people started automating the tasks. This process of automation started to generate a gamut of data. This generated data is collected, stored, and processed to extract non-trivial information by applying human wisdom.

This advent of technological advancements is generating data at a very high rate and this generated data needs to be collected and stored. Data is not only stored but needs to be processed to extract useful information from it. The processing of this huge amount of data also needs a high level of computing power. So, to have this high level of computing power, we have to expend huge resources in terms of manpower and finances. Often, neither individuals nor institutions have sufficient funds to procure the resources for the storage of data and for the computing power to process it. So, in the late 1960s ARPANET, which is the predecessor of the internet, came into existence and opened the gateway for sharing computing resources. In the 1970s and 1980s formed the seed for the idea of computer virtualization. This means running multiple operating systems on a single machine.

DOI: 10.1201/9781003138037-1

This technological development has introduced the concept of the client-server approach where a client can access data and applications available on the servers. In early 1995 the term, cloud started to appear in the network research community. With this came the concept of the Virtual Private Network (VPN). This revolutionary concept offered the same quality of services over the same infrastructure. Salesforce was the first organization to harness the potential of the sharing of resources over the internet via a website, named Salesforce.com.

Significant improvement of internet bandwidth by the service providers, fuelled ever increasing resource sharing over the internet. This gave birth to many organizations like Google, Netflix, Facebook, and the like. In the year 2006, Amazon launched its first web-based service, called AWS (Amazon Web Services), to provide storage and computing power to small business organizations over the cloud. This sparked competition between the industry giants to take advantage of the new technology of cloud computing (Marston et al. 2011; Voss 2010; Wang et al. 2010).

As the need to store and process data increases manyfold among organizations, so does the need for state-of-the-art technology (Khan et al. 2021). As these state-of-the-art technologies and infrastructure involve extra expenses, this naturally leads to the development of specialized designs of information technology infrastructure to cater for the demanding needs of small and medium enterprises (SMEs).

This delivery of various computing resources such as storage, processing power, and the like, over the internet via a server is referred to as cloud computing (Hayes 2008). Formally as per the National Institute of Standards and Technology (NIST):

> Cloud computing is a model for enabling convenient, on-demand network access to a shared pool of configurable computing resources (e.g., networks, servers, storage, applications and services) that can be rapidly provisioned and released with minimal management effort or service provider interaction.
>
> (Geelan 2009)

As the demand for computing resources has increased, the idea of on-demand services has emerged, thus making the organization, effectively no-software. (Grossman 2009) discussed different types of cloud: namely cloud that provides computing facility on demand and cloud that provides computing capacity. The former provides additional computing facility and the latter provides scalable computing power. An example of scalable computing capacity is Google MapReduce and Amazon Web Services is an example of scalable computing power.

In (Alam, Pandey, and Rautaray 2015), cloud computing is described more like distributed computing which facilitates the running of a program on many inter-connected computers simultaneously. Thus, their resources can be shared to achieve coherence and economies of scale. (Buyya, Ranjan, and Calheiros 2009; Khan et al. 2020) state that cloud is a virtualized, parallel, and distributed computing system to provide resources based on a service level agreement (SLA) that is agreed upon between the service provider and the consumer. The technology is not new, but it has been achieved through the journey from distributed through clustered to grid and so on.

The paper is organized as follows: **Section 1** discusses the architecture of cloud computing. **Section 2** explains the characteristics of cloud computing. **Section 3** presents Society 4.0. **Section 4** discusses the challenges of Industry 4.0. **Section 5** discusses Society 5.0. **Section 6** reviews some challenges of society 5.0. **Section 7** reflects on the comparative Google Trends in the advances discussed. **Section 8** concludes.

1.2 ARCHITECTURE OF CLOUD COMPUTING

A large pool of storage and/or computing resources that can be accessed via standard, defined protocols through an abstract interface is referred to as cloud computing. The cloud computing system can be categorized into two sections, referred to as front end and back end as shown in Figure 1.1. Both the front end and back end communicate with each other through the internet backbone. The foundation layer of the cloud contains raw hardware resources, components of which are mostly computing resources, storage resources, and network resources.

As shown in Figure 1.1, resources are virtualized to present them to upper layer and end users as integrated resources. These resources will be offered to the end user as: Infrastructure as a Service (IaaS). Above the infrastructure layer, there resides another layer referred to as the Platform layer which offers the cloud services as: Platform as a Service (PaaS). This layer adds the collection of specialized tools, middleware and services like operating system, programming interfaces, and the like. The next layer above the platform layer is the Software/Application layer which is referred in literature as Software as a Service (SaaS). This layer provides the end user with various business applications such as Customer Relationship Management (CRM), Web-services, for example.

1.3 CHARACTERISTICS OF CLOUD COMPUTING

The IaaS, PaaS, and SaaS features of cloud computing have important characteristics. These characteristics establish its similarity to and difference from traditional

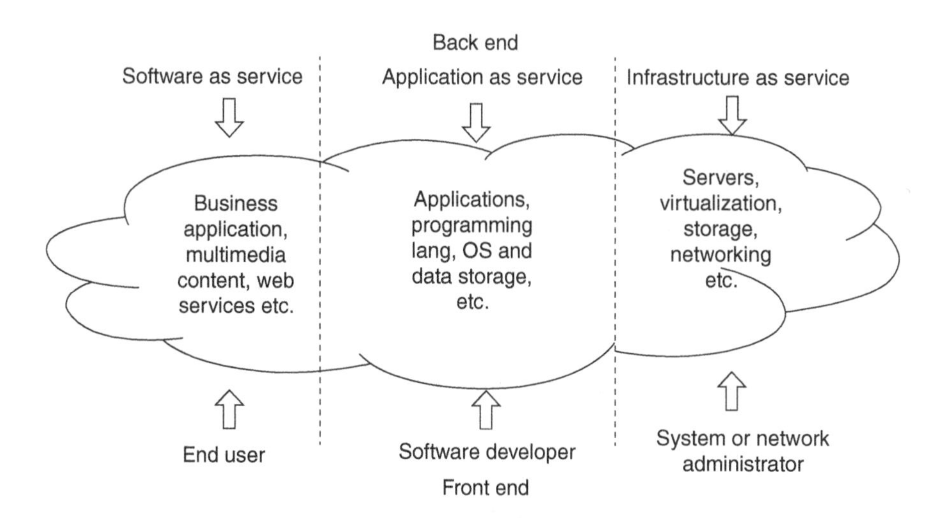

FIGURE 1.1 System architecture of cloud computing.

computing. According to NIST, there are five essential characteristics of cloud computing. The first of these characteristics is On-demand self-service. In this kind of provision, the user has control over whether to continue or to discontinue the provided services. Services provided in this way are mainly mail services, web services (like Amazon web services, IBM, Salesforce, and the like). The second characteristic identified by NIST is Broad Network Access. This provides network access to various clients (PDAs, mobile phones, laptops, and the like) through the normal mechanisms. The third characteristic mentioned by NIST is resource pooling. The computing resources are pooled to serve multiple consumers. These consumers might be using different kinds of physical and virtual resources to access the cloud. These resources can be dynamically assigned as per the requirements of the customers. The fourth characteristic is Rapid Elasticity. This means that resources can be automatically requested and released. This provision accommodates the consumer. The resources available for use are unlimited and can be scaled up or reduced by any amount, at any time. Measured Services are the fifth characteristic of cloud computing. The available resources are dynamically controlled and optimized by monitoring various aspects of the service (bandwidth, storage, active users, and so forth). This rapid development of information technology has driven the availability of information on the move. At the same time, it has a wide range of issues and challenges such as confidentiality, integrity, and availability, for example. Despite these challenges, information technology has strong penetration in society and has significantly revolutionized the industry.

1.4 INDUSTRY 4.0

The technological advancements have great impact on the human history of the industrial revolution and have been referred to as Industry 4.0 (Vogel-Heuser and Hess 2016). This is commonly known as the Fourth Industrial Revolution in human history. This claim refers to the current trend of automation, scaling and data exchange in industry. Cloud computing, artificial intelligence, the Internet of Things (IoT), and big data are integral parts of this industrial revolution (Prisecaru 2016). This Information Communication Technology is one the many trends that pushed industry to a new paradigm of:

- Shorter times to market because of the innovation and automation.
- Satisfying the customer with customization and higher product individualization.
- Flexibility to produce cost effective, small quantities of world class products, on demand.
- Decentralization of the hierarchies in the process of decision making.
- Increasing efficiency in terms of physical and economic cycles.

Thus, the technology evolution has strengthened the mechanization and automation, miniaturization, digitalization and networking, decentralization, development and production, and so forth (Lasi et al. 2014) (Nugent and Rhinard 2015; De España and SAU 2016). The fundamental concept of Industry 4.0 is based on the use of Cyber-physical system production and heterogenous data and knowledge integration, focusing on digitization, automation, optimization, human-machine interaction,

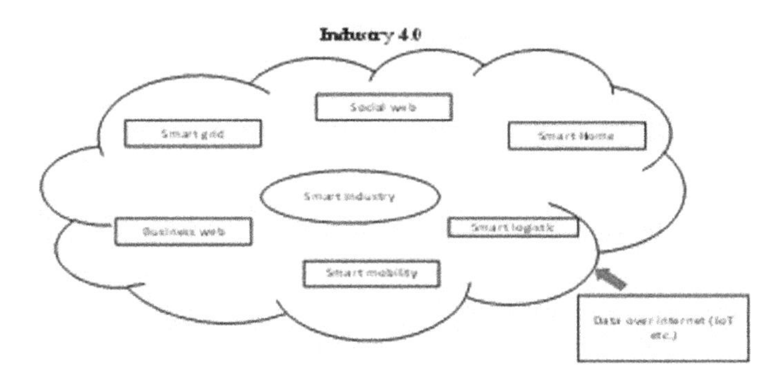

FIGURE 1.2 Industry 4.0.

automatic data exchange, and communication (Posada et al. 2015; Amit and Zott 2012; Casadesus-Masanell and Zhu 2013; Chesbrough 2007, 2010; Bucherer and Uckelmann 2011).

Figure 1.2 shows the systematic representation of the Industry 4.0 environment and its major components.

One of the advances of Industry 4.0 is in automating and configuring the equipment to adapt to the customer-specified manufacturing process. The impact of this would be to produce small and unique numbers of items. This might also facilitate in prototyping and producing new products quickly, in increasing the efficiency of production manyfold and in reducing the time to market significantly (European Union Chamber of Commerce in China 2017). Integrating product development with digital production also significantly reduces design, production, and delivery times. Sensor data reduces the error rate and critically monitors every detail of the production, increasing the efficiency and solving minor problems.

These advances in technological evolution has provided a plethora of analytical tools, the power of which can be harnessed to improve predictive maintenance, equipment failure, and hence downtime (Davies 2015).

1.5 INDUSTRY 4.0 CHALLENGES

Industry 4.0 is intended to be a major source of economic and social value creation, but at the same time it has some challenges to overcome (Davies 2015). These are: the investment costs of automation and information communication technology, collaboration with business organizations, service providers, and the like and the protection of the Confidentiality, Integrity, and Availability (CIA) triad in relation to large amounts of data collected from various sources. Also, it might have legal implications that need to be carefully addressed such as: staff supervision, product liability, and intellectual property rights, for example.

Some of the other barriers mentioned in literature are coordination among the organizational units, skilled manpower, cybersecurity issues when working with third party providers, the challenges of data ownership, concern about in-house activities, or about outsourcing the activities of the organization. One of the most challenging tasks is to integrate data originating from multiple sources with different architectures (Mckinsey and Company 2016; Khan and Kannapiran 2019). Improvements in technological advancement has improved human life and society as a whole. This perspective of improving upon social well-being is referred as Society 5.0.

1.6 SOCIETY 5.0

The advent of digitalization has created new values and is continuously becoming a vital pillar of industry and its policies. The concept of Society 5.0 was introduced in Japan in 2016 with the Japanese cabinet adopting the concept in its 5th Science and Technology Basic plan. The concept was identified as one of the major growth strategies for Japan. This idea was also termed as Super Smart Society. (Mouzakitis 2017) stated that the idea of progress is associated with modernity and social theory. So, using the basis of the current information society, in other words, Society 4.0, a new concept of human-centered prospering society was introduced as Society 5.0.

A high degree of convergence is achieved between virtual space and physical space by Society 5.0. Unlike Society 4.0, in Society 5.0, data is accessed via the cloud. In Society 5.0 a major source of data collection is by sensors in the physical space. This has minimized the gap between people, things, and systems, since all of them are connected. The results are extracted using Artificial Intelligence (AI) which is exceeding human capabilities and operating in physical space which was always the dream of the past. The concept is viewed from different perspectives by different nations depending on their situation. Besides addressing many social challenges faced and addressed by nations, Society 5.0 also has some challenges and issues associated with it.

1.7 SOCIETY 5.0: RISKS AND CHALLENGES

Along with creating numerous opportunities, Society 5.0 has some issues to be addressed. Digitization adds a major concern for society in controlling and handling the issues of cybersecurity. This is a serious concern and as the penetration of the internet into human lives increases, so does the risk that privacy and security might be compromised. Securing privacy in the cyber world is a major challenge (NISTEP 2019a, 2019b; Serpa et al. 2020). This privacy issue might cause serious damage to people's lives. Not only are breaches in personal data a matter of concern, but there could also be the risk of establishing a data elite. For the data elite, their data is stored and processed using cloud computing. Thus, power may shift to organizations who are not democratically elected, but to these data elite who control most of the data flow and storage.

Digital divide could be another issue for organizations and governments. As the reach of the internet into remote areas is still a dream. Not only the internet but its use will also be a major challenge. A good number of countries still lack digital infrastructure, and the internet is a dream yet to come true.

From the health point of view, this concept might cause digital fatigue, as has already been seen in the current pandemic of COVID. In addition to digital fatigue, human emotion could also be adversely affected as most of the communication might be with machines (Shahnawaz and Mishra 2015; Khan et al. 2018). This, in turn will degrade human emotions and capabilities (Kagermann and Nonaka 2019). This decline in human capabilities, consequently, might well lead to lower motivation. This could have long-term adverse effects on humans as it might result in humans who have become incapable of dealing with unanticipated events.

This approach might not only make humans less capable but also lead to a kind of society that is more divided. Humans could lead compartmentalized lives where they only know what they need to know and not beyond. The trait of human propensity for enquiry and expedition could degrade with time.

1.8 GOOGLE SEARCH TRENDS

Terms like digital transformation, cloud computing, Industry 4.0, and Society 5.0 have been extensively searched on the internet. Based on these global searches on Google search, data has been extracted and a graph has been plotted as shown in Figure 1.3 and Figure 1.4. Figure 1.3 shows the Google trend for the keywords: digital transformation, cloud computing, Industry 4.0, and Society 5.0. This trend shown here is from March 2020 to February 2021.

Figure 1.4 shows the google trend for the same keywords from March 2016 to February 2021

1.9 CONCLUSION

This study aimed to show the evolution of the digital trend from basic standalone computer applications to complex cloud computing via different stages of technological advancement. Through this process of digitalization it can be seen how industry got influenced by these innovations. This led to mechanization and

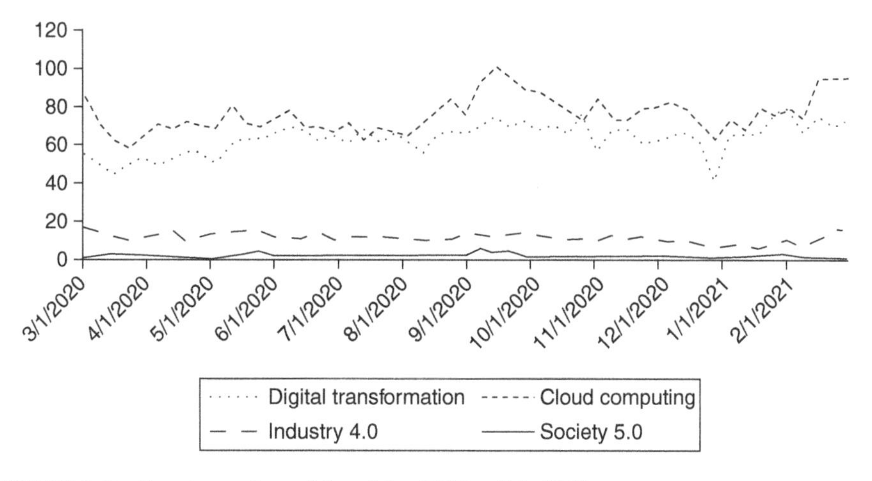

FIGURE 1.3 Google word trend from Mar. 2020 to Feb. 2021.

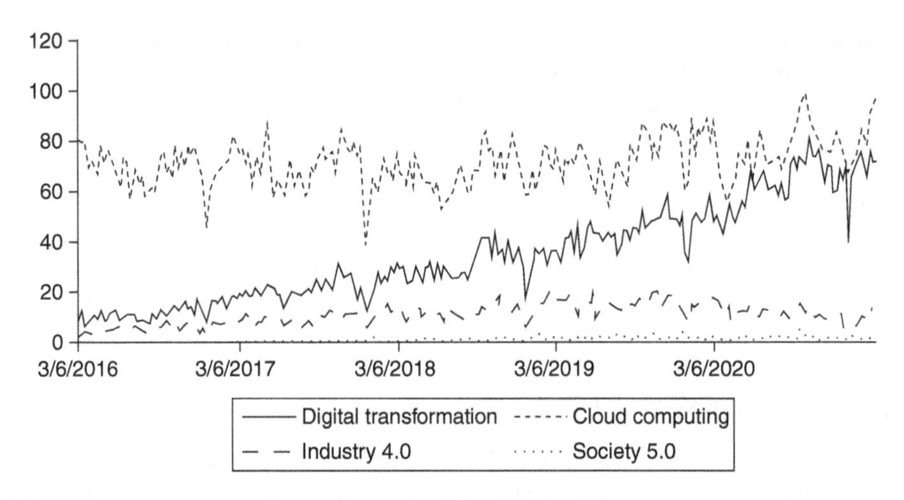

FIGURE 1.4 Google word trend from Mar. 2016 to Feb. 2021.

automation, miniaturization, digitalization and networking, decentralization, development, and production. Additionally, this industrial revolution has some issues such as coordination, data integrity, and the like. Digitalization has penetrated human society and makes human life more comfortable but has some issues and challenges to be addressed. Finally, Figure 1.3 and Figure 1.4 show how the terms: digital transformation, cloud computing, Industry 4.0, and Society 5.0 have become accepted and used worldwide.

REFERENCES

Alam, Md Imran, Manjusha Pandey, and Siddharth S. Rautaray. 2015. "A comprehensive survey on cloud computing", *International Journal of Information Technology and Computer Science*, 2: 68–79.

Amit, Raphael, and Christoph Zott. 2012. "Creating value through business model innovation", *2012*. In *MITSloan Management Review*, 53: 36–75.

Bucherer, Eva, and Dieter Uckelmann. 2011. "Business models for the internet of things." In Uckelmann D., Harrison M., Michahelles F. (eds) *Architecting the Internet of Things*. Springer, Berlin, Heidelberg. https://doi.org/10.1007/978-3-642-19157-2_10.

Buyya, Rajkumar, Rajiv Ranjan, and Rodrigo N. Calheiros. 2009. "Modeling and simulation of scalable cloud computing environments and the CloudSim toolkit: Challenges and opportunities." In *2009 International Conference on High Performance Computing & Simulation*, 1–11. IEEE.

Casadesus-Masanell, Ramon, and Feng Zhu. 2013. "Business model innovation and competitive imitation: The case of sponsor-based business models", *Strategic Management Journal*, 34: 464–82.

Chesbrough, Henry. 2007. "Business model innovation: it's not just about technology anymore", *Strategy & Leadership*, 35, 6: 12–17.

Chesbrough, Henry. 2010. "Business model innovation: opportunities and barriers", *Long Range Planning*, 43: 354–63.

Davies, Ron 2015 (R. EPRS | European Parliamentary Research Service. Members'| European Parliamentary Research Service. Members' Research Service PE 568.337).

De España, Gobierno, and SOCIEDAD UNIPERSONAL SAU. 2016. "Ministerio de empleo y seguridad social", *Estrategia de emprendimiento y empleo joven 2013/2016.*

European Union Chamber of Commerce in China 2017. China manufacturing 2025: Putting industrial policy ahead of market forces. //docs.dpaq.de/12007-european_chamber_cm2025-en.pdf

Geelan, Jeremy. 2009. "Twenty-one experts define cloud computing", *Cloud Computing Journal*, 4: 1–5.

Grossman, Robert L. 2009. "The case for cloud computing", *IT Professional*, 11: 23–27.

Hayes, Brian. 2008. "Cloud computing." In *Communications of the ACM*, 51: No 7. 9–11. ACM: New York, USA.

Kagermann, Henning, and Nonaka, Youichi. 2019. Revitalizing Human-Machine Interaction for the Advancement of Society: Perspectives from Germany and Japan. In Acatech DISCUSSION. Retrieved from https://en.acatech.de/wp-content/uploads/sites/6/2019/09/acatech_DISCUSSION_HumanMachineInteraction_final-1.pdf.

Khan, Shahnawaz, Ayman Al-Dmour, Vikram Bali, M. R. Rabbani, and Thirunavukkarasu Kannapiran. 2021. "Cloud computing based futuristic educational model for virtual learning." *Journal of Statistics and Management Systems*, 24: no. 2, 357–385. DOI: 10.1080/09720510.2021.1879468.

Khan, Shahnawaz, Usama Mir, Salam S. Shreem, and Sultan Alamri. 2018. "Translation divergence patterns handling in English to Urdu machine translation." *International Journal on Artificial Intelligence Tools*, 27, no. 05 (2018): 1850017.

Khan, Shahnawaz, and Thirunavukkarasu Kannapiran. 2019. "Indexing issues in spatial big data management." In *International Conference on Advances in Engineering Science Management & Technology (ICAESMT)-2019*, Uttaranchal University, Dehradun, India, 1–5.

Khan, Shahnawaz, Mohamed Redha Qader, Thirunavukkarasu Kannapiran, and Satheesh Abimannan. 2020. "Analysis of Business Intelligence Impact on Organizational Performance." In *2020 International Conference on Data Analytics for Business and Industry: Way Towards a Sustainable Economy* (ICDABI), 1–4. IEEE.

Lasi, Heiner, Peter Fettke, Hans-Georg Kemper, Thomas Feld, and Michael Hoffmann. 2014. "Industry 4.0", *Business & Information Systems Engineering*, 6: 239–42.

Marston, Sean, Zhi Li, Subhajyoti Bandyopadhyay, Juheng Zhang, and Anand Ghalsasi. 2011. "Cloud computing-The business perspective", *Decision Support Systems*, 51: 176–89.

Mckinsey&Company. 2016. *Industry 4.0 after the initial hype*, Mckinsey Digital.

Mouzakitis, Angelos. 2017. "Modernity and the idea of progress", *Frontiers in Sociology*, 2: 3.

NISTEP. 2019a. "Reasons why people feel anxious about the realization of Society 5.0 in Japan as of March 2019, by gender [Graph]", Retrieved May 10, 2020, from Statista website: www.statista.com/statistics/1058259/japan-negative-attitudes-society-5-by-gender.

NISTEP. 2019b. "Share of respondents who agree Society 5.0 will improve the quality of life in Japan as of March 2019, by gender [Graph]", Retrieved from Statista website: www.statista.com/statistics/1040691/japan-positive-attitudes-society-50-quality-of-life-by-gender.

Nugent, Neill, and Mark Rhinard. 2015. *The European Commission* (Macmillan International Higher Education, London).

Posada, Jorge, Carlos Toro, Iñigo Barandiaran, David Oyarzun, Didier Stricker, Raffaele De Amicis, Eduardo B Pinto, Peter Eisert, Jürgen Döllner, and Ivan Vallarino. 2015. "Visual computing as a key enabling technology for industrie 4.0 and industrial internet", *IEEE computer graphics and applications*, 35: 26–40.

Prisecaru, Petre. 2016. "Challenges of the fourth industrial revolution", *Knowledge Horizons. Economics*, 8: 57.

Serpa, Sandro, Carlos Miguel Ferreira, Maria José Sá, and Ana Isabel Santos. 2020. "Dissemination of Knowledge in the Digital Society". In Digital Society and Social Dynamics, 2–16. Digital Society and Social Dynamics. Stockport, Cheshire: Services for Science and Education, 2020. doi:https://doi.org/10.14738/eb.17.2020.

Khan, Shahnawaz, and Mishra, Ravi Bhushan. 2015. "An English to Urdu translation model based on CBR, ANN and translation rules." *International Journal of Advanced Intelligence Paradigms* 7, no. 1 (2015): 1–23.

Vogel-Heuser, Birgit, and Dieter Hess. 2016. "Guest editorial Industry 4.0-prerequisites and visions", *IEEE Transactions on Automation Science and Engineering*, 13: 411–13.

Voss, Adrian. 2010. "Cloud computing". Powerpoint slides, Hewlett-Packard Development Company, L.P. Retrieved from https://www.itapa.sk/data/att/628.pdf

Wang, Lizhe, Gregor Von Laszewski, Andrew Younge, Xi He, Marcel Kunze, Jie Tao, and Cheng Fu. 2010. "Cloud computing: a perspective study", *New Generation Computing*, 28: 137–46.

2 Cryptographic Algorithms and Protocols

Mohammad Khalid Imam Rahmani

CONTENTS

2.1 INTRODUCTION

In the rapidly growing digitization initiatives in the various Government and private departments, security of their valuable contents is a major issue. The information security assurance is key to winning the trust of users in the safety and secrecy of the data being shared by different parties over the internet or over any network channel (Bourgeois 2014). The economical availability of good quality communication

DOI: 10.1201/9781003138037-2

hardware and software tools has created tremendous opportunities for exploring more effective and efficient security techniques for securing the information of organizations (Soomro et al. 2016). Two contemporary technologies for the purpose are: cryptographic algorithms and cryptographic protocols (Gupta et al. 2016).

The reason why unauthorized parties become successful at reading secret information is that they have opportunities to access and reveal the secret information from secured systems (Bourgeois 2014) due to vulnerabilities in such systems. As a result, attackers can misuse or modify the information, reveal the secret information to dangerous parties, make wrongful representations to some organization, or make a plan for other harmful activities (Tsai and Chen 2013; Bashir et al. 2017). Cryptography provides a solution to this problem.

Cryptography uses cryptographic algorithms and protocols to make it difficult for any unauthorized users to reveal any restricted information (Mandal et al. 2012).

The main objective is to understand available tools and techniques and the importance of secure transmission of data while achieving authenticity, confidentiality and other security principles so that attacks can be prevented and secrecy of data can be ensured. Other objectives are:

1. To go through existing cryptographic techniques and to identify strong and weak points in the field of cryptography.
2. To have an insight into cryptographic algorithms and protocols.
3. To explore application areas of cryptography.

The cryptographic algorithms are described. The requirements of cryptographic protocols are discussed. Along with the conclusion, some application are.as of cryptography and research trends in information security have been explored.

2.2 PRELIMINARIES

Cryptography is an ancient Greek word in which 'crypt' means 'hidden' and 'graphy' means 'writing'. It is the science and art of attaining security by transforming original messages into unintelligible forms (Rosenheim 2020) or providing immunity against unauthorized access. Cryptographic algorithms are used to encode the messages before securely sharing the information through a network so that it becomes extremely tough for an unauthorized person to reveal secret details from the message.

The important components used in cryptography are summarized in Figure 2.1:

1. Plaintext and Ciphertext: The original message which the sender wants to share is called plaintext. At the sender's end, the plaintext is transformed into a secured form with an encryption algorithm. It is called ciphertext. At the receiver's end, a decryption algorithm is used to transform the ciphertext back into plaintext.
2. Cipher: The term cipher is used to refer to encryption and decryption algorithms. The cipher is used for different categories of algorithms in cryptography.

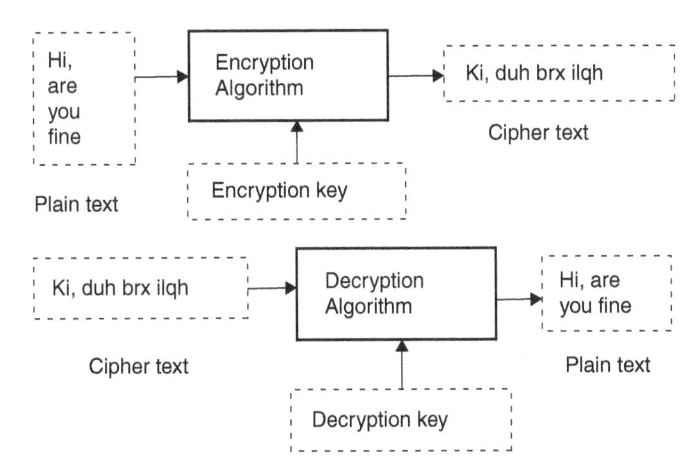

FIGURE 2.1 Basic Model of Cryptography. (Figure by the author.)

3. Key: A key refers to a number that an encryption/decryption algorithm uses to transform plaintext into ciphertext or vice-vera. To generate ciphertext from plaintext, an encryption key and an encryption algorithm are required. To obtain the original message back from the ciphertext, a decryption key and a decryption algorithm are required.
4. Alice, Bob and Eve: It is customary to understand three typical characters in cryptography who represent either computers or processes. Alice is the sender of secured data to the receiver Bob. Eve is the person who somehow intercepts the communication channel connecting Alice and Bob. Eve can either decipher the original message or she sends her own disguised messages to Bob.

2.2.1 Cryptographic Goals

Understanding cryptographic goals is essential for analyzing security issues for information systems, utilizing capabilities of cryptographic systems up to their full extents, and for measuring the strengths and weaknesses of cryptographic algorithms and protocols. There are four cryptographic goals described below:

- **Confidentiality:** is about ensuring access to information only for authorized parties. Confidentiality ensures privacy. There are many approaches to implement confidentiality such as: physical lock and key protection, passwords, and mathematical algorithms to make data unintelligible.
- **Data integrity:** is about safeguarding the information systems from any unauthorized changes to data. To ensure data integrity, the system must be able to detect any manipulation of data by unauthorized parties.
- **Authentication:** ensures the identification of parties trying to access data.
- **Nonrepudiation:** is a task that prevents a party from denying previous commitments or actions. If a dispute arises then third-party intervention is required to resolve it.

Any cryptography system must address all the four goals in practice (Stinson and Paterson 2018) because the objective of cryptography is to discover any unwanted trespassing and prevent its consequences such as, theft of information or any kind of fraudulent activities.

The most fundamental terms in cryptography are enciphering (encryption) and deciphering (decryption). Encryption transforms plaintext into ciphertext and decryption converts the ciphertext back into plaintext (Rosenheim 2020). A special number known as a key is used with the enciphering and deciphering processes.

2.3 CRYPTOGRAPHIC ALGORITHMS

Cryptographic algorithms are the set of mathematical and logical steps essential for transforming secret information into an encrypted cipher and for getting back the original information from the encrypted cipher. There are so many algorithms that are used in cryptography. The two encryption methods are symmetric key encryption and asymmetric key encryption.

2.3.1 Symmetric Key Algorithms

In symmetric key algorithms, the sender encrypts the plaintext and the receiver decrypts the ciphertext with a single secret key. They are also called secret key or private key algorithms. The key must be secured from unauthorized access because any party having the key can decrypt the sensitive data or even encrypt new data and then make the claim that it originated from the sender that was compromised. These algorithms are faster than asymmetric key algorithms and therefore they are used for larger data sizes. The shared key should be available only to the actual sender and receiver. The issue of key sharing between the sender and receiver causes many challenges. The diagram in Figure 2.2 shows the process of encryption/decryption for symmetric cryptography. The most common encryption techniques are described below:

2.3.1.1 Data Encryption Standard (DES)

This is the first encryption algorithm released by NIST. A team from IBM developed it in 1974. It was adopted as a national standard in 1997. It has both key size and block size of 64 bits. Only 56 bits of the 64 bits are used by the algorithm (Data Encryption

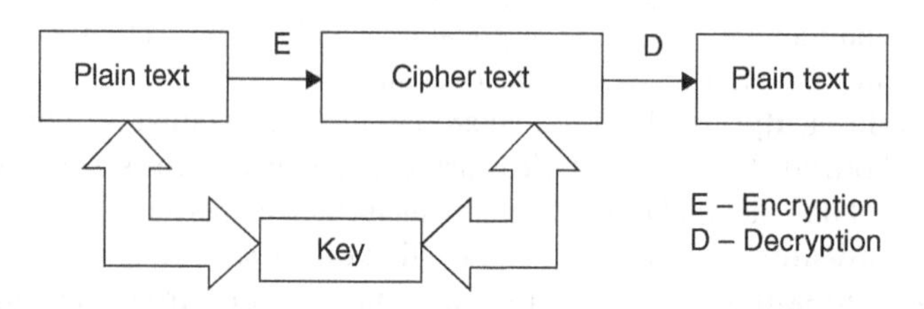

FIGURE 2.2 Symmetric Key Cryptography. (Figure by the author.)

Standard et al. 1999). The remaining 8 bits are set aside for the parity of other bits and are discarded later. DES uses a Feistel network. It served as a standard for securing secret commercial and unclassified data.

2.3.1.2 Triple DES or TDEA

This is an extension of DES. It has a key size of 192 bits with a block size of 64 bits. This encryption method differs from the original DES. It applies the DES cipher algorithm three times to enhance the level of encryption and the average safe time. TDEA is slower than other block cipher methods (Kelsey et al. 1996). As it is a strong encryption algorithm, it finds its application in the banking industry.

2.3.1.3 RC2

This uses 64 bit block ciphers with key sizes from 8 to 128 bits. It uses 18 rounds of two different types called MIXING (16 rounds) and MASHING (2 rounds).

Blowfish: uses a 64 bit block cipher. It was designed as a replacement for the DES algorithm. It applies key sizes from 32 to 448 bits. Blowfish takes 14 or less rounds.

Advanced Encryption Standard (AES): is a symmetric key encryption/decryption algorithm. It is a block cipher method. It supports a block size of 128 bits and key sizes of 128, 192, or 256 bits. Its default is 256 bits. Joan Daemen and Vincent Rijmen developed it to become the winners of a competition conducted by NIST to replace DES. Consequently, the US Government adopted AES, superseding DES. AES is a special case of the Rijndael algorithm which can select block/key sizes of 128, 160, 192, 224, or 256 bits. NIST published it as FIPS 197 on 26th November 2001. AES Standards are summarized in Table 2.1 (Dworkin et al. 2001). In the case of 128 bits key length, the number of rounds is 10 (9 processing rounds and 1 extra round performed at the end of the encipher stage). In the case of 192 bits key size, the number of rounds is 12. In the case of 256 bits key length, the number of rounds is 14. Encryptions performed by AES (Zhang et al. 2021) are fast and flexible. It is suitable for different platforms. This algorithm uses a substitution-permutation network. Its performance is good in terms of both software and hardware. AES uses a Non-Feistel network.

TABLE 2.1
Summary of AES Standards

AES Standards	Key Size (in bits)	Block Size (in bits)	Number of Rounds
AES-128	128	128	10
AES-192	192	128	12
AES-256	256	128	14

Each round is carried out in four steps:

1. Substitute bytes: It makes a non-linear substitution of one byte with another according to a lookup table. It ensures non-linearity.
2. Shift rows: It makes a transposition of the last three rows which are shifted cyclically for a certain number of steps.
3. Mix columns: A linear operation carried out on the columns of the state. It combines the four bytes of each column. Shift rows and mix columns provide diffusion.
4. Add Round key: Each byte of a subkey is combined with the corresponding byte of the state.

2.3.2 Asymmetric Key Algorithms

In asymmetric key algorithms, separate keys are used to encrypt and decrypt the data. One key is the public key used for encryption which must be shared with the sender. Another key (private key), used for decrypting, must be kept secret. Therefore, these are also known as public-key algorithms. Asymmetric encryption algorithms like RSA (Zhang et al. 2021) cannot encrypt large amounts of data. The diagram in Figure 2.3 illustrates the mechanism of encryption/decryption in public key algorithms:

The messages can be encrypted with both the public key and the private key. For decryption, only the private key can be used. These encryption systems ensure the goal of confidentiality because a message encrypted by any sender using the receiver's public key can only be decrypted by the receiver's paired private key. In digital signature schemes of public-key cryptography sender authentication (Sharma and Singh 2021), integrity and nonrepudiation are ensured (Forouzan 2011). Asymmetric algorithms are slower, but they do not face the issue of key distribution. Examples of asymmetric algorithms are Diffie-Hellman, RSA, and DSA. Rivest, Shamir, and Adleman (RSA) is one of the popular public key algorithms used for encryption purposes and for digital signatures.

2.3.3 Digital Signature Algorithm

Digital Signature Algorithm (DSA) is also a public key algorithm that is used only for digitally signing documents. This scheme is suitable for achieving authentication before a message or documents are shared (Forouzan 2011). Receiving a digitally

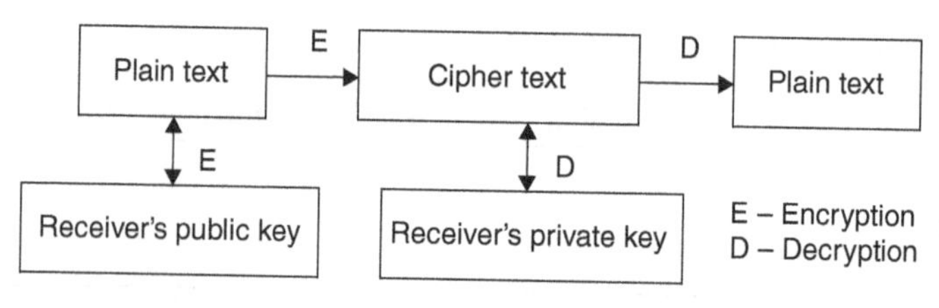

FIGURE 2.3 Asymmetric Key Cryptography. (Figure by the author.)

signed document, the recipient becomes confident that the sender is genuine and the document has not been altered during the transmission. Digital signatures are applied in software distribution, financial transactions, and for documents that might be tampered with. To verify the document, the receiver performs the following steps:

1. Decrypts the digital signature using the sender's public key to read the message.
2. Generates a message digest for the receiver's message using the same algorithm used by the sender.
3. If both message digests do not match, the sender's message digest is compromised.

2.3.4 Hash Functions

Hash functions, or one-way functions, are used in public-key cryptography for implementing protocols (Alawida et al. 2021). Hash functions do not need any key. They are easily computable but harder to reverse. For example, f(x) can be computed easily but the computation of x from f(x) would take many years even if all the world's computers worked on it collectively. The value of f(x) is a fixed-length hash value computed from x which is the plaintext. Neither the contents of the plaintext nor its length can be obtained. Hash functions are used to verify the integrity of the documents and encryption of passwords. Even a small change in the contents can be easily detected because the hash values of the two versions will be significantly different.

2.4 CRYPTOGRAPHIC PROTOCOL

Cryptography analyses the issues of integrity, authentication, privacy, and nonrepudiation. Cryptographic algorithms are academic importance (Schneier 2007). Application of these algorithms alone cannot guarantee achievement of the goal of cryptography. Well-defined policies and agreements between the parties involved in the communication are also required, to make cryptography a reliable technology for achieving its goals so that it can solve real problems in completing online tasks between trusted parties.

A cryptographic protocol is a distributed algorithm designed to precisely describe the interactions between two or more parties with the objective of implementing certain security policies. It follows a series of steps in exact sequence. Every step must be completely executed without any alteration in the agreed-upon sequence. It must be complete and able to finish a task. At least two parties are required. Any single party executing a series of steps to complete a task is not a protocol. Every party must know, understand, and follow it. They must not be able to do something beyond the specified agreement.

A cryptographic protocol uses a cryptographic algorithm to achieve the goal.

2.4.1 Arbitrated Protocols

Arbitrated protocols use a trusted third party, called an arbitrator. The arbitrator has no vested interest and cannot favor any of the involved parties. Such protocols are used to complete tasks between two or more parties not trusting each other.

2.4.2 Adjudicated Protocols

The arbitrated protocols are implemented with two subprotocols to reduce the cost of third-party involvement. A non-arbitrated protocol is used in the first instance which is executed for each task. If necessary, a second level, an arbitrated protocol, is used which is executed only in case of disputes that occur between the involved parties during the task.

2.4.3 Self-Enforcing Protocols

These protocols require no arbitrator to complete tasks or to resolve disputes. The protocol itself ensures that there is no dispute between the involved parties. One party can detect if the other party diverts from the protocol and the task is stopped immediately. It is ideal that every protocol should be self-enforcing.

Similarly, to attacks on cryptographic algorithms and techniques, protocols can also be attacked by a criminal.

2.4.4 Types of Protocols

2.4.4.1 Key Exchange Protocols

A key exchange protocol is required for two parties to reach an agreement for a shared secret key. Either party can authenticate the other or both parties can authenticate each other. The protocol can agree on the generation of a random key. One party can generate the key and send it to the other party or both parties can participate in the key generation.

2.4.4.2 Diffie-Hellman Key Exchange

This protocol is used by the involved parties to agree on a shared key by exchanging messages through a public channel. Therefore, the key is not revealed to any unauthorized party. This provides protection only against passive attacks.

2.4.4.3 Identification and Authentication Protocols

Identification protocols are required to ensure the identity of both parties when they are online for a task. Genuine possession of their private keys needs to be verified. The level of identification by the protocols may be judged with three levels:

1. Who is he? – Biometrics is used.
2. What he possesses? – Hardware devices are used.
3. What he knows? – Secret keys or passwords are used.

Some of the popular protocols used are: zero-knowledge protocol, Schnorr Protocol, Guillou-Quisquater protocol, witness hiding identification protocols, and the like.

2.4.4.4 Using Password Authentication

In absence of any digital signature scheme, the two parties can share a password. This is comparatively less powerful.

2.4.4.5 Protocol Using Digital Signatures

Digital signature-based protocols are used to protect against active attacks by authenticating the two parties.

2.5 ISSUES IN CRYPTOGRAPHY

In symmetric cryptography, if the key is lost, communication cannot be completed. This creates an issue around secure key distribution by possibly needing either the sender and the receiver to communicate directly, needing to operate via a trusted third party, or needing to communicate via an existing cryptographic medium (Sharma et al. 2021; Khan and Kannapiran 2019). The issue of key distribution is to be dealt with delicately: keys must be stored and used, as well as destroyed, securely.

Cryptography only transforms plaintext but never hides it (Rahmani et al. 2014). One weakness of cryptography is, if somehow any third party detects the presence of an encrypted message, it can make attempts to break into it out of curiosity. Sometimes curiosity initiates the attempt to break the code. Consequently, the secrecy can be overcome and the information modified or misused.

2.6 CONCLUSION

For a secret communication, secrecy of messages must be ensured. In this chapter, a short account of the techniques and mechanisms for information security for sharing secret information between two or more parties, is provided. A detailed description of both cryptographic algorithms and protocols is given.

Future works in the field need to be selected to explore useful techniques that can enhance the security of information and enhance the ease and confidence of sharing secret information online. Securing a secret message is the primary goal. A study of cryptanalysis is also required to test the information security systems with more advanced cipher breaking techniques in an authentic operating environment. Additionally, there is the need to develop an information security infrastructure framework with modern cryptographic tools and techniques that will save time and increase the capacity of hidden secret messages for sharing confidential information with online trusted parties.

REFERENCES

Alawida, Moatsum, Azman Samsudin, Nancy Alajarmeh, Je Sen Teh, Musheer Ahmad, and Wafa' Hamdan Alshoura, et al. 2021. "A novel hash function based on a chaotic sponge and dna sequence," *IEEE Access*, 9: 17882–17897.

Bashir, Tariq, Imran Usman, Shahnawaz Khan, and Junaid Ur Rehman. 2017. "Intelligent reorganized discrete cosine transform for reduced reference image quality assessment," *Turkish Johurnal of Electrical Engineering & Computer Sciences*, 25, no. 4: 2660–2673.

Bourgeois, David. 2014. *Information Systems for Business and Beyond*. The Saylor Foundation, Washington, DC.

Data Encryption Standard et al. 1999. "Data encryption standard". In *Federal Information Processing Standards Publication*. p. 112.

Dworkin, Morris J., Elaine B. Barker, James R. Nechvatal, James Foti, Lawrence E. Bassham, E. Roback, and James F. Dray Jr. 2001. Advanced Encryption Standard (AES), *Federal Inf. Process. Stds.* (NIST FIPS), National Institute of Standards and Technology, Gaithersburg, MD, [online], doi.org/10.6028/NIST.FIPS.197 (Accessed March 28, 2021).

Forouzan, Behrouz A. 2011. *Cryptography and Network Security*, 2nd Edition, Publisher McGraw-Hill Education (India) Pvt Limited, New Delhi.

Gupta, Brij, Dharma P. Agrawal, and Shingo Yamaguchi. 2016. *Handbook of Research on Modern Cryptographic Solutions for Computer and Cyber Security*. IGI Global, Hershey, PA, USA.

Khan, Shahnawaz, and Thirunavukkarasu Kannapiran. 2019. "Indexing issues in spatial big data management." In *International Conference on Advances in Engineering Science Management & Technology (ICAESMT)-2019*, Uttaranchal University, Dehradun, India.

Kelsey, John, Bruce Schneier, and David Wagner. 1996. "Key-schedule cryptanalysis of idea, g-des, gost, safer, and triple-des". In *Annual International Cryptology Conference*. Springer, Manhattan, NY, USA pp. 237–251.

Mandal, Akash Kumar, Chandra Parakash, and Archana Tiwari. 2012. "Performance evaluation of cryptographic algorithms: DES and AES". In: *2012 IEEE Students' Conference on Electrical, Electronics and Computer Science*. IEEE, pp. 1–5.

Rahmani, Md. Khalid Imam, Kamiya Arora, and Naina Pal. 2014. "A Crypto-Steganography: A Survey," *International Journal of Advanced Computer Science and Applications*, 5, no. 7: 149–155.

Rosenheim, Shawn James. 2020. *The Cryptographic Imagination: Secret Writing from Edgar Poe to the Internet*. JHU Press, Baltimore, MD.

Schneier, Bruce. 2007. *Applied Cryptography: Protocols, Algorithms, and Source Code in C*, John Wiley & Sons, Hoboken, NJ.

Sharma, Shivam, Jain Sajal, and Chandavarkar, B. R. 2021. "Nonce: Life cycle, issues and challenges in cryptography," In *ICCCE 2020*, Springer, Manhattan, NY, USA pp. 183–195.

Sharma, Deepak, and Avtar Singh. 2021. "Privacy preserving on searchable encrypted data in cloud," In Gurdeep Singh Hura, Ashutosh Kumar Singh, and Lau Siong Hoe (Eds.), *Advances in Communication and Computational Technology*, Springer, Singapore, pp. 847–863.

Soomro, Zahoor Ahmed, Mahmood Hussain Shah, and Javed Ahmed. 2016. "Information security management needs more holistic approach: A literature review". *International Journal of Information Management*, 36, no. 2: 215–225.

Stinson, Douglas Robert and Maura Paterson. 2018. *Cryptography: Theory and Practice*. CRC Press.

Tsai, Ming-Hong and Chaur-Chin Chen. 2013. "A study on secret image sharing". In *Proceedings of the 6th International Workshop on Image Media Quality and its Applications*, Citeseer, Tokyo, Japan, pp. 135–139.

Zhang, X. Hu, J., Li, H. Guan. 2021. "A comprehensive test framework for cryptographic accelerators in the cloud," In *Journal of Systems Architecture*, 113: 101873.

3 How to Enhance Data Privacy on Android

A Proposal

Bharavi Mishra, Aman Ahmad Ansari, and Poonam Gera

CONTENTS

DOI: 10.1201/9781003138037-3

3.1 INTRODUCTION

Android is a trending and popular smartphone and tablet operating system (OS) developed by Android Inc., which was later acquired by Google in the year 2005. IDC's 2020 report states that there were 1.3 billion Android-based mobile phones, roughly 84.8% of the overall smartphone market (IDC 2020) Android is based on Linux Kernel version 2.6 and is extensively modified to use sophisticated software and hardware sensors to bring innovation and value to customers. Android devices also contain popular web applications such as Gmail, YouTube, Facebook, Instagram, and the like, which make it more efficient and useful in everyday usage.

Smartphones, tablets, and smart TVs store a large amount of users' data (Khan and Kannapiran 2019). If this data were to be revealed, it would be a critical threat to the privacy of the user. Unfortunately, the popularity of Android smartphones has also attracted malware developers. A study (Kaspersky 2014) performed by Kaspersky revealed that there were over 10 million dubious apps on the Google Play Store, that might send users' sensitive data (such as photos, call logs, etc.) via the public network to an unknown destination without the users' knowledge. Therefore, the default permission given to applications to access user data over the internet is more cumbersome to protect users' privacy (Bashir et al. 2017). Although the Marshmallow and higher version updates offer users the option to choose which set of data they are permitted to access, it is only available to about a hundred variants of smartphone. It takes only 7.5% of the total Android market.

The last decade has witnessed numerous occasions where data has been stolen from the phones and other digital devices of many important people. One such case appeared when Australian Foreign Minister, Julie Bishop's phone was hacked on her foreign visit after the Malaysia Airlines flight MH17 incident (*Daily Mail* 2014). These instances present a threat to the security of a nation as well as the person's privacy. Hence, to ensure the security and privacy of the user, Android OS should have a robust security mechanism (Android 2020).

The most common and standard mechanism of data security is encryption. It renders the data useless unless the key is available to decrypt the data. This method has proven effective in the past; however, this mechanism is computer-intensive on resource-constrained android devices.

An alternative technique designed by the security community to preserve the user's privacy is through detecting an app as malware using static and dynamic analysis and to deny it a place in the play store. Many researchers have created and reinforced similar systems (IMPACT 2018; Tracedroid 2012; Weichselbaum et al. 2014; Droidbox 2019) to observe the behavior of the application while it is executing on an emulator. In these systems, an application is restricted to running the application in an emulated environment. They statistically analyzed an app based on specific heuristics and further classified it as benign or malware. Furthermore, Android developers designed Google Bouncer (Trend Micro "A look a Google Bouncer") based on the above concept. It examines all applications uploaded on the Google Play Store using a QEMU-based emulator (Oberheide and Miller 2012). Consequently, Google Bouncer has contributed towards a reduction in the amount of malware downloaded from Google Play by 40% in the year 2011 (Lockheimer 2011).

Emulator identification is a crucial step for any malware, and there has been extensive research conducted to differentiate real devices from emulators (Jing et al. 2014; Vidas and Christin 2014; Petsas et al. 2014). Therefore, these existing approaches accomplished the task efficiently when they were initially developed. Eventually, the malware developers discovered vulnerabilities to evade the detection by emulator-based analysis tools. For example, applying static analysis on APK will not yield any result if the application package (APK) file is encrypted (Broadcom 2012). Moreover, if the application successfully identifies the emulator, then the dynamic analysis technique can be evaded. If the application executes on an emulator, then there is a high probability that it is being run in a sandbox. If the emulator is identified by the malware, it acts as if benign, just like any other normal application, and reduces the chances of being caught in the analysis.

This chapter is an extension of our previous work, submitted as a student paper to: The LNM Institute of Information Technology. In this chapter, we will discuss one such approach (developed and tested by our team) to protect user privacy. This approach performs behavioral analysis of Android applications. It combines two independent systems, based on client-server architecture, to secure users' data residing on their smartphones.

The server helps to understand the application's actual behavior and tags it appropriately. Simultaneously, based on the application's behavior, the client (SP-Enhancer) secures the data transmitted across the internet (sent by the application under consideration) from the smartphone. The server incorporates the properties of a real device to perform emulation-based malware detection. The primary focus of this environment is to simulate user behavior such that the malware doesn't detect the emulator and reveal its real characteristics. After behavioral analysis, if the application is genuine, SP-Enhancer preserves privacy by converting the private or sensitive data not required by the application into an unreadable format by scrambling it randomly without using any encryption. The traditional methods such as Brute-force, Dictionary Attack, or Rainbow Table Attack are not able to descramble the scrambled data.

3.2 ADVANCES IN PRIVACY PRESERVATION AND SECURITY

Since the inception of mobile computing, new technologies have developed rapidly. Security experts are now focusing on strengthening the security and the privacy of mobile devices. In this section, the existing tools and techniques for mobile data security are discussed. These methods are either developed by Android developers or by researchers. The security solution provided by Google runs all the applications on the emulator to identify any existing security flaws before uploading it to the Google Play Store. Malware applications sidestep these solutions because they can identify the emulators using the emulators' behavioral analysis. These applications show benign behavior on such emulators. This section will discuss four basic techniques for preserving privacy and security: Data Encryption methods, Android Security Model, Third Party Techniques, and Emulator detection and bypassing.

3.2.1 Data Encryption Method

Encryption is an encoding technique, which was developed to protect data from unauthorized access. It secures the data but the price of this is the cost of encrypting/decrypting. Due to this, the performance of the system and the user experience suffers drastically. Therefore, Android versions are not using encryption (Seppala 2015), which further increases the security threat to the user data.

3.2.2 Security Models Used by Google

Google implemented some security protocols in Android architecture to provide security and privacy to the user. In this section, we will discuss some of the security mechanisms used by Android.

3.2.2.1 Application Manifest

Developers define the application properties (Android, "App Manifestl Android Developers") in the application manifest, which provides the necessary information to execute any application (Yuksel, Zaim, and Aydin 2014). The developer digitally signs every Android application. After that, the Android's security model maps the developer's signature with the unique ID of the application package to enforce signature-level permission authorization (Android, "Security Tips"). This security model only confirms the origin and provides integrity protection of the source code.

A new update of an application can bring dangerous permissions without the user's authority (Hoffman 2014). To handle this issue and make it more user-friendly, Google divided the permissions into thirteen different permission groups. During the application update, the application can acquire new permissions within the acquired permission group, which poses an additional threat to data security and privacy (xda-developers.com "Play Store Permissions Change Opens Door to Rogue Apps"). All the applications also have default permission to access the internet since most of them require the internet for proper functioning. *Access of internet* permission belongs to the *other* permission group (Hoffman 2014).

3.2.2.2 Sandboxing

Similar to Linux, in Android, each application is installed with a unique user-id (UID), which consequently results in a different home directory for every application, where the app's code and data reside. These directories are only accessible through the process running with the corresponding UID or within the application's user space. This mechanism secures the application's personal or local data from all other applications.

Just like the other security mechanisms in place, the application's sandboxing is efficient but not invincible (source.android.com "Application Sandbox") because applications with the necessary permissions can access these directories.

3.2.2.3 Google Bouncer

This scans Android applications to detect potentially malicious applications without disrupting the user experience. This service performs a set of analyses on all the applications against known malware such as Trojans and Spyware. It compares the application's behavior with the formerly analyzed applications and tries to detect any red flags.

However, according to research (Trend Micro 2012), it was found that Bouncer can be fingerprinted. It is not so difficult for the application to take advantage of this vulnerability and conceal its actual behavior to masquerade as a legitimate app when running on Bouncer. Application scanning only takes five minutes on Bouncer, which is not sufficient for most of the malicious applications. Any malicious application can act as a benign one while being tested and might do severe damage subsequently. Moreover, the malicious application can also perform an update-attack, wherein there is no malicious code in the initial installer payload. However, when the application performs its update, it will install the malicious code on the device (Trend Micro 2012).

3.2.2.4 Malware Removal

Android is designed in such a way as to be able to detect malware and protect the platform against any potential threat. Therefore, if any device is affected, the Play Store can remotely remove most of the malware from any Android device (Android.com "Application Sandbox").

Unfortunately, malware removal is possible only if we can detect the malicious application in time. Although it is possible to remove installed applications remotely, it would be better to protect the user device from the beginning (Trend Micro 2012).

3.2.3 Third-Party Applications

Apart from the default built-in methods, there exist various third-party techniques to help with security and privacy preservation. We will discuss a few of them in this section:

3.2.3.1 TaintDroid

TaintDroid monitors how third-party applications utilize the private data of the smartphone user. It uses a dynamic taint tracking and analysis system. When an application is installed, the user is prompted with the permissions required by the application to function, but it does not show how the application will use the data. TaintDroid takes a pragmatic approach towards making the users aware of how applications are using their data. It uses the technique of *Dynamic Taint Analysis* and is capable of tracking data such as GPS, phone number, IMEI, and the like. TaintDroid provides real-time analysis of applications (Enck et al. 2014; Appanalysis.org "Realtime Privacy Monitoring on Smartphones"; Techrepublic.org "TaintDroid: Warns about Android Apps Leaking Sensitive Data"). Dynamic Taint Analysis is the ability to monitor program code as it runs on the system (Schwartz, Avgerinos, and Brumley 2010). TaintDroid marks the relevant information for the user and tracks it as it flows in Android. It taints the information at the source and then tracks it at the sink (where the information leaves the device) (Techrepublic "TaintDroid: Warns about Android Apps Leaking Sensitive Data").

3.2.3.2 LP-Guardian

LP-Guardian focuses on maintaining users' location privacy. LP-Guardian provides tracking, profiling, and identification of threats while maintaining full application functionality. The framework first intercepts the location API, called from the application, and then scrambles the data retrieved so that malicious applications cannot identify the users' actual locations (Fawaz and Shin 2014).

Although emulator and dynamic analysis-based systems gain popularity towards controlling the invasion of different applications in devices, at the same time, enormous work has been done in the field of emulator detection and dynamic analysis bypassing. Now the important work that has taken place, relating to the proposed framework, will be discussed. Jing et al.(2014) identified more than 10 000 heuristics to detect Android emulators and ranked the top 30 artifacts of detection heuristics with their respective accuracy. These heuristics were further subdivided into file, API, and property of emulator. This study was done for QEMU and Virtualbox-based emulators and also developed an Android application to detect whether a device is a real device or an emulator.

Vidas and Christin (2014) have detected emulator systems using behavioral analyses such as system performance, working of hardware and software components, and system design choices to bypass emulator-based Android malware analysis environments. They have also focused on CPU and GPU performance differences along with other hardware and software components.

Petsas et al. (2014) presented an anti-analysis technique. This technique is based on static information such as device IMEI, IMSI, SIM Number, dynamic information such as sensor specific readings, and VM-related Android emulator properties to identify and bypass emulator-based Android analysis environments.

Maier et al. (2015) fingerprinted ten different sandboxes. They demonstrated that dynamic code loading could bypass Google Bouncer. The study suggests that malicious applications use dynamic code loading more frequently in comparison to benign applications and concluded that neither the static nor the dynamic analysis could provide comprehensive security from malware if the attacker designed the application to behave differently in different scenarios (real or emulated).

Although the Android based system has too many security consolidators, there are still some breaches in ensuring the privacy and security of Android devices. TaintDroid only reveals the type and the source of the data leaked. LP-Guardian, on the other hand, secures only the location of the user's device. Existing tools certainly provide a level of privacy, but not comprehensive privacy for all user data. To provide a complete solution, the applications and the data cannot be addressed with one generic solution.

Further, existing tools use encryption to attain privacy in data sets. However, encryption is a traditional technique, which can be broken through Cold-Boot Attack, Brute Force Attack, etc., and is computer-intensive in nature (Trend Micro "Application Sandbox"). On the other hand, existing emulator-based systems are easily detectable and can be bypassed by malware applications.

Therefore, we will discuss a consolidated environment in which a QEMU-based emulator is used along with static and dynamic analysis. It also uses a scrambling technique to protect user privacy against unauthorized non-functional data requirements.

3.3 SECURITY IN ANDROID

3.3.1 Android Architecture

Android is a Linux-based open-source software program or, rather, an Operating System (OS) (see Figure 3.1). The platform security is based on the OS's architecture, which is achieved by separating resources and accessibility into subsequent layers. The Linux kernel runs on the lowest level. Each layer assumes that the layer below is secure. Subsequent layers become less and less accessible.

The Android low-level security model is based on application sandboxing (Android "Android Security 2017 Year in Review"; Android "Android Security 2018 Year in Review"). Android sandboxing is the process of isolating an application in the system. It prevents outside influences on the layers mentioned in the architecture. All applications are assigned user IDs while they are running. They have access rights to their own files only. It prevents outside malware and security threats since, if an application experiences a security problem, other applications' operations will not be affected.

Android provides hardware-backed protection of keys for cryptographic services. The stored keys provide a safe, secure channel for authentication of user data. Verified Boot is used to check the state of the system when it starts ("Android Security 2017 Year in Review" 2017). It verifies whether the system is in a good state. In Android 8.0 (Oreo), Google introduced Project Treble to increase low-level security(Android "Android Security 2017 Year in Review"). Project Treble separates the Open-Source Android OS framework from the hardware code implementations at the vendor level. It has had a positive impact on device security and the speed of updates.

In earlier versions, device manufacturers and System on Chip vendors had to update a major chunk of the Android code if they wanted to update their distributions as there was no separation between the Android OS and vendor hardware. Now, the OS can be updated without changing or re-configuring hardware implementations. It also provides an advantage in the Hardware Abstraction Layer (HAL). In older versions of Android, HALs were made to run in-process. Instead of having HALs in-process, HALs have now been isolated to their own processes. This promotes the principle of least privileges, as HALs in a process do not have access to the identical set of permissions compared to the rest of the process (Android "Android Architecture").

3.3.2 Application Security

Android uses the permission model to prevent an app from using sensitive data and resources which are not required. Apps need corresponding permissions to use application programming interfaces (APIs) to interact with the underlying system (Zhou et al. 2011; Ongtang et al. 2012; Bugiel, Heuser, and Sadeghi 2013; Backes et al. 2014). All permissions taken by an app must be specified in the application's manifest file (Karim, Kagdi, and Di Penta 2016). Permissions are grouped into three categories corresponding to the risk and security level associated with resources and APIs: *normal, dangerous, and signature permissions. Normal permissions* include the permissions where the application has to interact with resources out of the sandbox and do not pose any threat to user privacy. Normal permissions include Bluetooth, internet, and KILL_BACKGROUND_PROCESS. *Signature permissions* are granted at the install time and allow an application to use the permissions signed by the identical certificate. *Dangerous permissions* are the permissions that could pose a potential threat to security and privacy. The user is required to approve these permissions. SMS, storage, and Camera permissions belong to this category.

In earlier versions (until Android 5.0), users could not choose a subset of permissions. They had to grant all the permissions mentioned in the application's manifest file during its installation. Android 6.0 (Android Marshmallow) introduced a new mechanism for permission, called runtime permission (Android "Android Platform"). Where users are notified of the dangerous permissions on runtime and can choose to withdraw or deny any specific permission, it requires ample knowledge

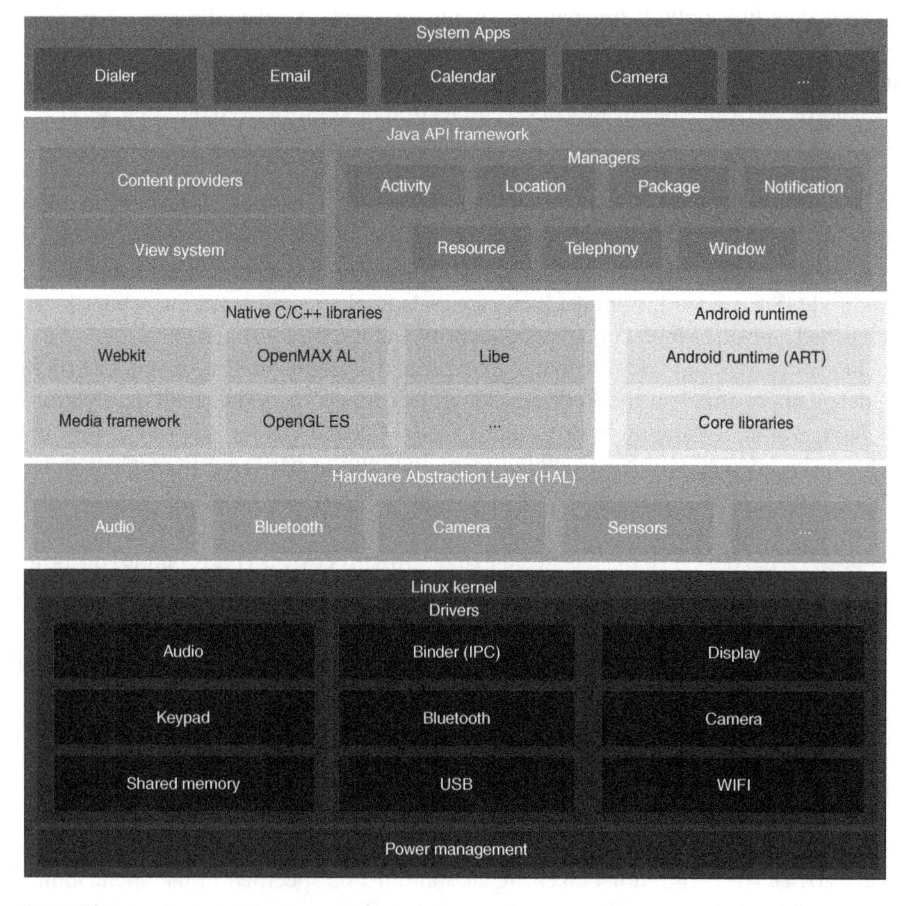

FIGURE 3.1 Android Platform Architecture ("Platform Architecture" 2021).

of the application and domain to decide which permission/s is/are essential. However, sometimes, denying permissions may cause malfunctioning. Therefore, it is essential to know what set of permission is mandatory to run any application.

Android also provides strict policies for sensitive APIs. In Android 8.0 and above, the GET_ACCOUNTS permission is now no longer sufficient to gain complete access to the list of accounts active on the device. For example, the user must now grant permission to the Gmail app to access the Google account on the device even though Google owns Gmail. As an example, *Settings.Secure.ANDROID_ID or SSAID* is an ID provided to all apps. To prevent misuse of *ANDROID_ID*, Android 8.0 provides a mechanism that does not allow any change to *ANDROID_ID* when the application is re-installed until the package name and key are identical. Another feature, *Build.getSerial()* returns the actual serial number of the device until the caller has the PHONE permission. Android 8.0 has made this a deprecated API. This protects the serial number of the device from being misused by the applications.

Android has seen advances in toughening security policies. However, it is worth noting that, as of October 2020, only 40.35% of devices are running Android 10.0, and 22.59% of devices are running Android 9.0 (Pie). More than 35 percent of users are still using older Android versions on their phones (Statcounter "Mobile & Tablet Android Version Market Share Worldwide"). Due to a lack of user knowledge and security breaches in sensitive APIs, users are often manipulated.

3.4 PROTECTION FRAMEWORK

The protection framework uses client-server architecture where the server has sufficient computing power and clients are Android devices. This framework preserves data privacy utilizing a two-step process, performed at both ends.

After initiating the app installation, the client requests the server to check the app. The server responds to the request performing the requested app safety analysis through static and dynamic analysis. If the application is malware, the server prompts the client not to install the app, and if the app is already installed, it instructs the client to remove the app immediately. In contrast, if the app is benign, all requested permissions by the app may not be genuine. Therefore, the server performs the next step by scanning the requested permission with respect to the app's genuine requirements.

Genuine permissions with respect to apps are identified and stored on the server in *Server-App-List*. *Server-App-List* is a text file generated by the permission recommender system at the server end, using machine learning techniques. Further, this information is communicated to the client, which updates its *Client-App-List*. Using the *Client-App-List*, the client module only sends genuine data in the original format and other non-genuine data in a scrambled format. If, somehow, a malware app bypasses this security check, user data privacy is still preserved, because by default, everything will be sent in scrambled form. All these tasks are performed at the client-side using the SP-Enhancer framework, which comprises four components: *Recommender-App*, *Client-APP-List*, *Internal-Routine*, and *Scrambler*. The block diagram and control flow of the complete framework is shown in Figure 3.2.

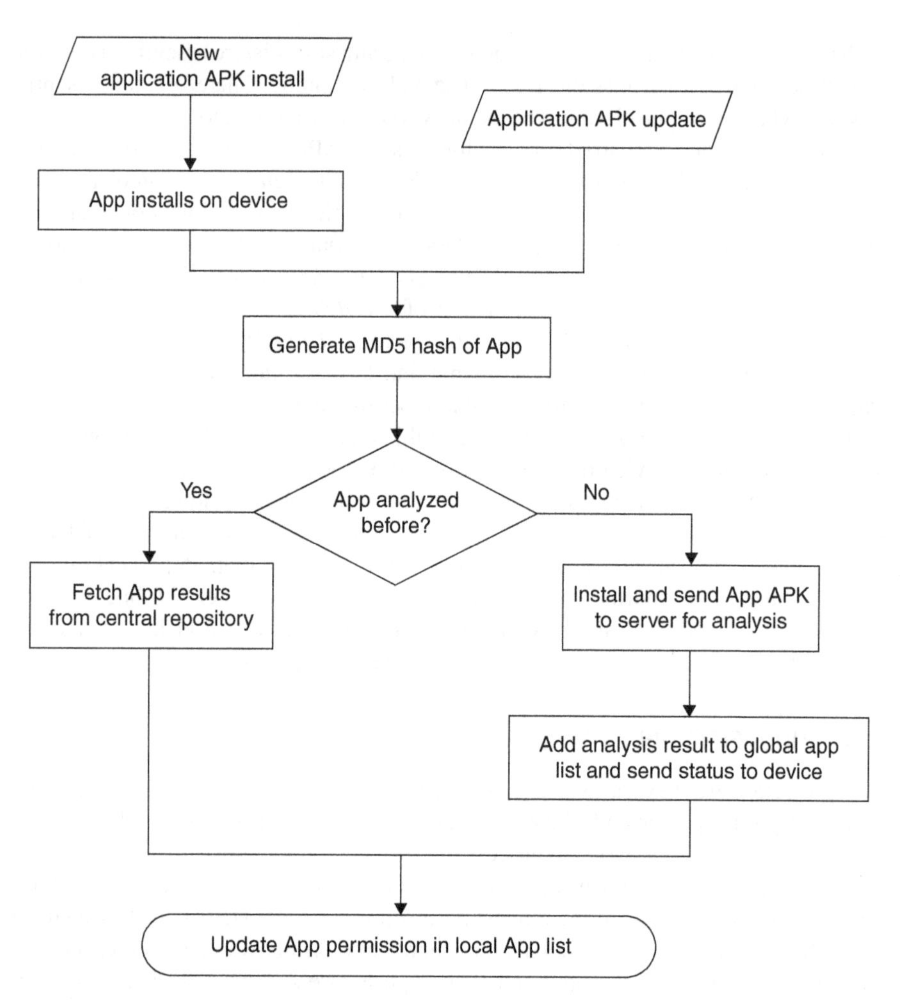

FIGURE 3.2 Block Diagram of Framework.

3.4.1 Functions of the Server

The server scans all the apps requested by the client to preserve the security and privacy of the client's data from unauthorized access. This has three major components: *Notification-List*, *Server-App-List*, and *APP-Analysis-Environment*. After receiving the client's request, the server first makes checks. If the app is already present in the list, it immediately responds to the client with the list of genuine permissions corresponding to the app. Otherwise, all the security check requests along with the client id and app id, are stored in *Notification-List*. After a detailed security check, if the app is benign, genuine permissions associated with that app are stored in *Server-App-List*. This information is also communicated to all the clients listed in *Notification-List* corresponding to the APP id.

3.4.1.1 Notification List

Notification-List is dynamic in nature and serves a book-keeping function for the clients who want to install a new application. It is a mapping of Android applications and clients who request to use it. It stores the information in the form of array of lists, where the first element of each list stores the APP id, and the remaining elements of the lists store clients' ids who initiated the security check. After the security check, *Server-App-List* is updated. All the clients associated with the APP are notified, and the corresponding list is destroyed.

3.4.1.2 Server App List

It stores the outcome of the APP's security check performed by the server. Each row of this list contains a hash value of the app's id, corresponding tags: malware or benign, a list of genuine permissions, in case of a benign app, and time of analysis. When a new request comes from the client, the requested app's hash value is searched in this list. If the matches, the application tag and permission list are sent to the client in the case of a benign application; otherwise, only the tag value is sent. However, if there is no match, the app id is added to the *Notification-List*, and security analysis of the requested app is performed. After that, the *Server-App-List* is updated, and the outcome is also communicated to all the clients listed in the *Notification-List.*

Having a *Server-App-List* substantially reduces the repetitive analysis of the same application. Once the behavior is recognized, all the previously registered clients are also notified about the new findings.

Hash(APP Id)	Application_tag	Permission_list	Time

3.4.1.3 App Analysis Environment

This performs security and privacy analysis of all the clients' apps. The analysis of all these apps is performed on an emulator. The emulator exhibits real device behavior to understand the true behavior of apps, that will be shown after the installation. This task is performed in two steps. The first step, defined as *Permission-Scanner*, dynamically scans each app's behavior on a real emulator and sends the analysis result to *Permission-Recommender*, if the app is not malware; otherwise, it terminates at this step. After that, *Permission_Recommender* will identify the necessary or genuine set of permissions using the application's categories defined by the Google Play Store and stored in *Server-App-List.*

3.4.1.4 Permission Scanner

This module has two sub-components: *Static-Analyzer* and *Dynamic-Analyzer.* Initially, the *Static-Analyzer* performs the static analysis of the application package (APK) file. It returns the information about the permissions required and the set of activities and services associated with the application. However, if the APK is encrypted, static analysis is not possible. On the other hand, *Dynamic-Analyzer* runs on an emulator controlled by the intelligent environment so that any malware won't detect the emulator.

TABLE 3.1
Permission Groups

Permission Types	Description
Location	Location-related permissions
External_Storage	Permission(s) to Read/Write
Network_State	Permission(s) to check network state
Phone	Permission(s) related to call log
Contact	E-mail, mobile numbers, etc.
Calendar	Read or write user's calendars
SMS	SMS permissions
Hardware	Sensors and sensors' related permissions

- *Static_Analyzer:* This uses Androguard (Google. "Androguard"), a reverse engineering toolbox, to extract package information from the APK files. Initially, *Static_Analyzer* uses *AndroapkInf* to extracts the application's manifest file from the APK and stores it in different files. The manifest file contains a list of all the permissions required by the application and the activities, services, broadcast receivers, and any native code. The permissions gathered from the manifest file are grouped together based on their data requirements. For instance, *ACCESS_FINE_LOCATION* and *ACCESS_COARSE_LOCATION* permissions are placed together as they both need GPS sensor data. The permission groups used in the framework are listed in Table 3.1. A linear sweep over APKs can obtain static features. Androguard (static analyzer) extracts static features from applications' APK files. Each app can be represented using a feature vector. Each feature vector contains information about permissions, packages, hardware, intents, and class (i.e., malware or benign). Thereafter, any classification algorithm can be used to predict malware apps.
- *Dynamic_ Analyzer*: Droidbox (Github. "Droidbox: Dynamic Analysis of Android Apps") along with some additional add-ons and Morpheous (Jing et al. 2014) testing base, are used in *Dynamic_Analyzer*. It comprises of four sub-modules (Personalizer, Sensor Simulator, Event Simulator, and User Simulator) to convert the emulator into a fake or real device to observe Android applications' real behaviors.
  - *Personalizer*: All real devices, such as mobiles, tablets, and the like, have user phone logs, contacts, messages, files, pictures, and so forth. An emulator, by default, doesn't contain any of these. Personalizer overcomes this gap by populating the emulator with fake user content. To perform this, Personalizer takes the *Permission.txt* (the output of the *Static_Analyzer*) file as the input and populates the call logs, contacts, storage, and so on, based on the permissions requested by the application. For the fake user content, it uses open-source datasets. Even though the phone numbers are randomly generated, the contact names are genuine. As an initialization of the emulator, if the application requests access to memory, the Personalizer

also populates the *SD* card with random files having bogus data: characters, songs, images, etc. Nevertheless, if the APK file is encrypted, all these activities will automatically occur, although it is computer-intensive because we must populate every type of data.

- *Sensor-Simulator*: Every real device comes with embedded sensors and a mechanism to access its data. However, in the emulator, the sensor-related data is static, which might be detected by most APPs. Therefore, Sensor-simulator feeds the sensors with bogus data. It uses *Permission.txt* to get all the sensors' permissions required by the application and accordingly feed the manipulated data in real time delays to make the emulator realistic. Unfortunately, few sensors, such as Accelerometer, can be used by any app without permission. Therefore, it may cause a threat to *Dynamic_Analyzer*. Thus, the emulator is fed with bogus data taken from open-source, where the subjects are walking, driving, running, and the like. Similarly, Sensor Simulator also feeds data to other relevant sensors from anonymous contributors (Android "Set up Android Emulator Networking").
- *Event-Simulator*: Some applications need the device's broadcast messages. Event-Simulator's task is to create events according to the application's requirements. It uses *Receivers.txt* (received from *Static_Analyzer*) to know which broadcast is required by the application. Event-Simulator constitutes the scripts corresponding to every type of broadcast intent. For instance, the most distinct feature of any emulator is battery level. It is always consistent, with a value of 50%, unless it is altered. Malware can detect emulators by merely monitoring the battery level. Hence, the Event Simulator tells the emulator to change the battery level by updating its status based on the dataset (Rotterdamopendata "Rotterdam Open Data").
- *User-Simulator:* User actions and gestures are one of the main components of any Android device, which are not available in the emulator. User-Simulator bridges this gap by simulating fake user behavior on the emulator. The user interface (UI) is either developed in Java or native code. Java-based UI is designed for all general-purpose applications. However, developers tend to prefer native code if the application is resource-intensive because it is executed on the CPU, unlike JAVA code, which is executed on the Dalvik Virtual Machine or Android Runtime (OpensourceForU "What A Native Developer Should Know About Android Security?"). Due to these properties, User-Simulator is divided into two parts: JAVA Based-User Simulator and Native Code Based-User Simulator. The following sections discuss them in further detail.
- *Java-Based-User-Simulator*: It fetches UI components using *UI Automator* (Android "UI Automator"). The application's extracted activities are stored in an XML file named "UI dump". This file is analyzed using the XML parser, and components are separated, based on their attributes: such as clickable, focusable, and long-clickable, and so on. This information is given to Monkeyrunner ("Android "Monkeyrunner") script to simulate artificial user behavior. The framework selects a random UI element to perform a specific task. However, we can only simulate a limited amount of UI elements'

activities. Therefore, an event such as 'scrollable' is simulated using 'touch' and 'drag' activities. 'Long-clickable' action is triggered with 'touch' and 'hold' gestures, followed by a 'drag' event. Similarly, the element's location on display is identified by the 'bound' attribute in [X, Y] [X_Length, Y_Length] format. All the triggered events are targeted at the middle of the UI element to achieve maximum accuracy.

- *Native-Code-Based-User-Simulator*: Gaming applications require intensive computing power and graphics; therefore, they are developed in native code to enhance the throughput of the overall system. A probability-based approach is used to retrieve the UI elements of gaming applications. We use a pre-computed probability distribution matrix of the UI elements. Using this technique, we increase the probability of analyzing most of the application features instead of random input of activities. The minimum size of a button is 48 dp ("Material.io "Metrics & Keylines") on a UI. We have separated the UI screen into cells of 48dp x 48dp matrix. The following formula is used to convert dp to absolute pixels:

$$pixels = \frac{\text{dp} * \text{Display\_Metrics}}{\text{Default\_Density}} \tag{1}$$

Where display metrics is defined as the density of height or width of an emulator device. *Default_Density* is 160 (standard reference density) (Android "DisplayMetrics"). The screen pixel matrix is computed using equation (1). After calculating the matrix, we have considered the top 50 free gaming applications available on the Google Play Store and calculated any UI element's existence. The cells are then sorted in descending order, based on their probability. The elements on the sorted list are then given as an input to the Monkeyrunner (Google "Androguard") script to simulate input with a latency (response time taken by the Android activity after the event was triggered) of one second. The following equations calculate the probability:

$$Q(\boldsymbol{i}, \boldsymbol{j}, \boldsymbol{k}) = \begin{Bmatrix} 1, \textit{if there is button on } i^{th} \textit{ row and } j^{th} \textit{ column} \\ 0, \textit{otherwise} \end{Bmatrix} \tag{2}$$

$$P\left(i, j\right) = \sum_{k=0}^{50} Q(i, j, k) / 50 \tag{3}$$

Where Q(i,j,k) represents the probability that a button exists in the kth activity. P(i, j) is the actual probability of the button existence in the cell of ith row and j^{th} column in the screen matrix. Rather than giving pseudo-random inputs to Monkeyrunner (Android "UI/Application Exerciser Monkey"; AGoogle "Androguard") script, the framework provides input based on probability, which is systematic and reduces unnecessary random input computations.

While executing the app in a simulated environment, we can extract behavioral features such as File I/O, Network I/O, program trace, and so on. This information is used to train the classification model to predict malware or benign apps.

3.4.1.5 Permission Recommender

Even though an app is benign, all the requested permissions may not be genuine. Therefore, this module suggests the set of permissions essential for a particular app. It performs the same operation, for each of the apps of the Google Play Store, categorizing them and identifying the minimum set of permissions required by each category. The permissions are also divided into two different clusters: dangerous and standard permissions. The necessary set of permissions for a specific app is determined by comparing the minimum set of permissions required by the category to which the app belongs, and the total permissions required by the app. This helps to quantify all dangerous permissions requested by the app that are not essential for the app's functioning and which compromise the user's privacy. This task is performed in three steps: Data Collection, Data Analysis, and building the User-Facing Recommender APP (SP-Enhancer). Using this app, a user can enable/disable permissions they have already granted/not granted.

3.4.1.5.1 Data Collection

In this step, we have collected information about top-rated apps (that is, those with a rating >4 on a scale of 1 - 5) in each category available at the Google Play Store in one Google Sheet. These apps were grouped into 40 categories using the Google Play Store's categories. An app category is extracted from the span[itemprop=genre] element of the document object returned by the Jsoup connection. A Jsoup connection is initiated by providing the Google Play Store webpage of a particular app. The google sheets API provides a clean interface to interact with a Google Sheet.

3.4.1.5.2 Data Analysis

After collecting the app permissions data, a recommendation matrix is created. Market basket analysis is used to create this matrix. In this approach, the similarities and the differences between the minimum set of permissions for the category to which the app belongs, and the total permissions required by the app is used to identify the **Essential and Non-Essential Permissions.**

- **Essential and Non-Essential Permissions**. After categorizing the apps, the permissions are grouped into essential and non-essential permissions. Permission is essential if most of the apps in a category require that permission for functioning and other permissions are non-essential permissions. Non-essential permission is not always used for malicious activities. For example, an app may use SMS access permission for automatic OTP verification even though most of the apps in that category may not require automatic OTP verification, so they might not need SMS data. However, we treated SMS access as non-essential permission for that category because app developers may scrape all our SMS messages into their database to perform user behavioral analysis.

3.4.1.5.3 *Recommendation Matrix Preparation*

The apps belonging to a particular category should request only the permissions required by the majority of apps in that category. This scheme ensures that every app contributes to deciding the essential and non-essential permissions. This information is stored in a two-dimensional matrix, known as the permission recommendation matrix, where rows represent apps and columns represent permissions. In this matrix, the digit, 1 in the ith row and j^{th} column show that the ith app needs the jth permission, and the digit, 0 shows that the ith app does not require the jth permission. The number of times that the digit, 1 occurs in a column indicates the importance of that permission. This number and the statistical measures of mean and median, help to demarcate essential and non-essential permissions. Experiments with both mean and median separately, concluded that the mean gives better results in identifying non-essential permissions. The permission whose mean value is less than the user-defined threshold is considered as a non-essential permission. In this framework, this value is 0.5.

3.4.2 Client Side

On the client-side, **SP-Enhancer** is used to ensure privacy. Whenever the client installs any application, the framework first searches the Hash(APP Id) value in the *Server-App-List*, if it's available. The corresponding tagging (malware or benign) along with the permission in case of when it's benign, is sent to the client. Otherwise, the app is sent for scrutiny. *The SP-Enhancer framework* comprises four components: SP-Enhancer (Android Application), an app list, an Internal-Routine (Background Service), and the scrambler module. A high-level diagram of SP-Enhancer is shown in Figure 3.3. The application creates and updates the client app list file that contains all the installed applications. The internal routine works in sync with the server and updates its app list file with the appropriate application permission. The scrambler scrambles the data, based on the application's permissions stored in its app list.

3.4.2.1 Recommender App

It helps the user to decide which permission is essential for proper functioning. When a user installs any app, the action *android.intent.action.PACKAGE_INSTALL* is triggered.

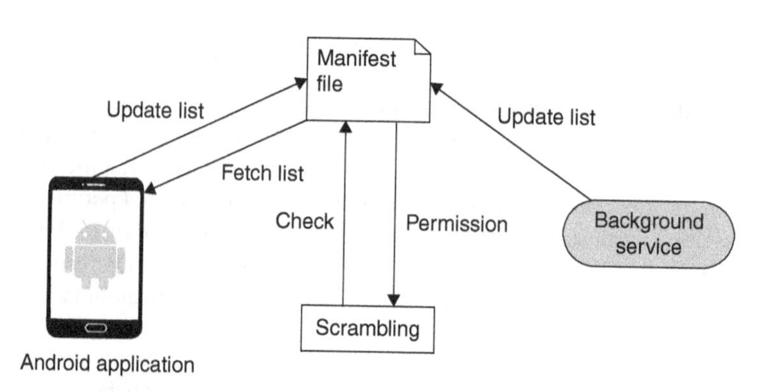

FIGURE 3.3 SP-Enhancer: Data Flow between Components (Singh, Mishra, and Gera 2015).

On the receiver's Receive(), we extract the package name from the intent received and check the Google Sheet API to get the permissions that should be granted. An activity-open event is triggered, and Recommender Activity opens up, which fetches the data from the recommendation matrix. At this time, there are three possible scenarios:

- If the app information is stored in the recommendation matrix, corresponding essential and non-essential permissions are shown to the user.
- If the app doesn't request any permission (there are all 0's in the rows corresponding to the app), the framework shows that the app is safe to use.
- If the app is not in the database, checks are made in respect of its category and users are informed about essential permissions.

3.4.2.2 App List

The app list file contains applications' UIDs, package names, permissions to send data, and the hashes of the APKs. When any application initiates data transfer over the internet, firstly App_List is checked for permission. If the permission is granted, then the packets are constructed, and data transfer occurs; otherwise, the original data is scrambled and then put into the packets for communication. The application's hash value is used to check its analysis result on the server.

The app list file resides in the internal directory of the application. In Android, all applications run in their own process space to protect their data from unauthorized access. However, some applications can get access to it. Hence, to secure app list from unauthorized access, a lightweight encryption is performed. When the list is created, it will be encrypted using a key generated by performing an XOR operation on the SP-Enhancer's UID and a random number (generated at runtime). The random number and the list will be kept in the process space of the SP-Enhancer application, from where the scrambler can access it. The copy residing in the application directory will be encrypted and will only be decrypted using the reverse process.

3.4.2.3 Internal Routine

The internal routine is responsible for communication between the client and server. Whenever a new application is installed on the smartphone or for the existing applications, the internal routine will calculate the new application's hash value and send it to the server for analysis and update the app list. After the analysis, it further updates the app list according to the server's information.

- **Scrambler:** This core component's task is to transfer non-essential data in the bogus format. As mentioned earlier, Android is based on the Linux kernel, which resides in the bottom-most layer of the Android OS stack. Like Linux, all processes are descendants of the "*init*" process, which is the first to be called during boot time. All processes are directly or indirectly called by the "*init*" process, including the Zygote process. When the Zygote process initiates, it initializes the Virtual Machine (VM), completes loading the library, and initializes operation (Hu and Zhao 2014). Whenever the Zygote fetches a new application process, a VM instance is copied to the new application process and assigns a new distinct VM instance to the application (Android "Zygote").

```
struct task_struct{
...
...
/* task state */
pid_tpid;
pid_ttgid;
...
/* process credentials */
char comm[TASK_COMM_LEN];   /* executable name */
...
...
};
```

FIGURE 3.4 Struct task_struct.

Inside the Linux kernel, every application is allocated a unique UID, PID and GID (Android "User IDs and File Access"). The permissions are assigned to a specific UID rather than the application. When two or more activities of the same application have multiple processes, their PID, as well as their GID, are different, but the UID remains the same because they are being executed in the same process space and belong to the same application.

Kernel space is structured such that each process is allocated a *task_struct* data structure, shown in Figure 3.4, which stores the information about the process such as the name of the process, its PID, UID, GID, permissions, and so forth, (Android "Kernel/Goldfish"). The current running application thread can be accessed using this structure in the kernel space using the name "*current*" (returning a pointer to the currently running *task_struct*). This pointer to the structure is crucial. It helps to find the thread feature that transmits the data over the network. All the applications either use TCP or UDP for data transmission; therefore, data should be scrambled at the transport layer. The scrambler fetches the PID and the UID of the process/thread using the *current* pointer. Once the UID is retrieved, it scans through the app list to get the application's permissions. The permission flags in the app list file control the scrambling process. If the permission flag is the digit, 0, data is sent without tampering. However, if the flag is the digit, 1 or the application UID search does not give any result, scrambled data is transferred. All the data scrambling should be done before checksum calculation; otherwise, the Android structure will not send the modified data. The sequence diagram showing the communication between different components of the SP-Enhancer framework is depicted in Figure 3.5.

During data transmission, the system's efficacy is of prime importance; therefore, the scrambler scrambles only 10% of the data.

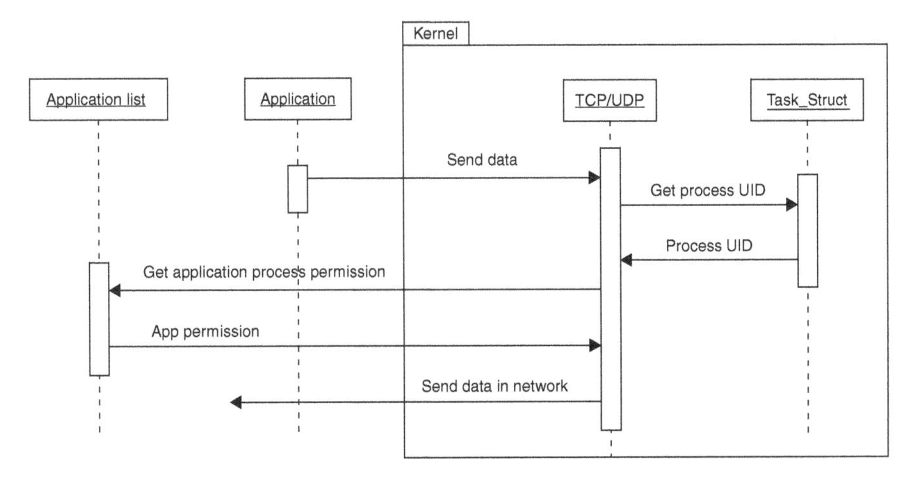

FIGURE 3.5 Sequence diagram of the SP-Enhancer framework (Singh, Mishra, and Gera 2015).

3.5 CONCLUSION

Android is the most popular operating system available on smartphones and many other devices, including watches and televisions. Its vast application has various security mechanisms, but it fails to provide protection against data theft by malicious applications. A new and efficient framework to secure sensitive data from unauthorized access has been discussed. The framework benefits in terms of the number of computations needed as it does not require any cryptographic techniques to protect the data.

This framework uses a self-learning system to detect malicious apps and secure the users' smartphone data. SP-Enhancer sends users' sensitive data (both multimedia and text) in a scrambled form as per the app list information. This framework works for both the technologically-aware and the opposite.

ACKNOWLEDGMENT

The authors would like to thank Shirish Singh for his valuable time and contributions.

REFERENCES

Android. 2012. "Kernel/Goldfish." android.googlesource.com/kernel/goldfish/+/android-goldfish-2.6.29/include/linux/sched.h.

Android. 2012. "Zygote." groups.google.com/g/comp.mobile.android/c/fxRSAv8t2p8?pli=1.

Android. 2015. "App Manifestl Android Developers". Accessed April 21, 2015. developer.android.com/guide/ topics/ manifest/ manifest-intro.html.

Android. 2017. "Android Platform." developer.android.com/about.

Android. 2017. "Android Security 2017 Year in Review."

Android. 2017. "Android Security 2018 Year in Review." 2018.

Android. 2020. "Android Architecture." source.android.com/devices/architecture.
Android, 2020 "Monkeyrunner." developer.android.com/studio/test/monkeyrunner.
Android. 2020. "Security." source.android.com/security.
Android. 2020. "Security Tips." developer.android.com/training/articles/security-tips.
Android. 2020. "Set up Android Emulator Networking." developer.android.com/studio/run/emulator-networking.
Android. 2020. "UI/Application Exerciser Monkey." developer.android.com/studio/test/monkey.
Android. 2020. "UI Automator." developer.android.com/training/testing/ui-automator.
Android. 2021. "Application Sandbox." source.android.com/security/app-sandbox.
Android. 2021. "DisplayMetrics." developer.android.com/reference/android/util/DisplayMetrics.html.
Android. 2021. "Platform Architecture." 2021. 2021. developer.android.com/guide/platform/index.html.
Android. 2021. "User IDs and File Access." developer.android.com/guide/topics/permissions/overview.
Appanalysis.org. 2014. "Realtime Privacy Monitoring on Smartphones." 2014. www.appanalysis.org/.
Backes, Michael, Sven Bugiel, Sebastian Gerling, and Philipp Von Styp-Rekowsky. 2014. "Android Security Framework: Extensible Multi-Layered Access Control on Android." In *Proceedings of the 30th Annual Computer Security Applications Conference*, 46–55. ACM, New York, NY, USA.
Bashir, Tariq, Imran Usman, Shahnawaz Khan, and Junaid Ur Rehman. 2017. "Intelligent reorganized discrete cosine transform for reduced reference image quality assessment." *Turkish Journal of Electrical Engineering & Computer Sciences* 25 (4): 2660–2673.
Broadcom 2012. "Obfuscating Embedded Malware on Android." https://bit.ly/3xGL5ic.
Bugiel, Sven, Stephan Heuser, and Ahmad-Reza Sadeghi. 2013. "Flexible and Fine-Grained Mandatory Access Control on Android for Diverse Security and Privacy Policies." In *Proceedings of the 22nd USENIX Conference on Security*, 131–46. USENIX Association, Washington, DC.
BusinessWire, 2016. "Smartphone Growth Expected to Drop to Single Digits in 2016, Led by China's Transition from Developing to Mature Market, According to IDC." www.businesswire.com/news/home/20160303005418/en/Smartphone-Growth-Expected-to-Drop-to-Single-Digits-in-2016-Led-by-Chinas-Transition-from-Developing-to-Mature-Market-According-to-IDC.
Daily Mail. 2014. "Foreign Minister Julie Bishop's Phone HACKED on Foreign Soil While She Was Dealing with MH17 Tragedy." www.*dailymail*.co.uk/news/article-2726227/Revealed-Foreign-Minister-Julie-Bishops-phone-HACKED-foreign-soil-dealing-MH17-tragedy.html.
Enck, William, Peter Gilbert, Seungyeop Han, Vasant Tendulkar, Byung Gon Chun, Landon P. Cox, Jaeyeon Jung, Patrick McDaniel, and Anmol N. Sheth. 2014. "TaintDroid: An Information-Flow Tracking System for Realtime Privacy Monitoring on Smartphones." *ACM Transactions on Computer Systems* 32 (2): 1–29. doi.org/10.1145/2619091.
Fawaz, Kassem, and Kang G Shin. 2014. "Location Privacy Protection for Smartphone Users." In *Proceedings of the 2014 ACM SIGSAC Conference on Computer and Communications Security*, 239–250. ACM, New York, NY, USA.
Github. 2019. "Droidbox: Dynamic Analysis of Android Apps." github.com/pjlantz/droidbox.
Google. 2013. "Androguard." code.google.com/archive/p/androguard/.
Hoffman, Chris. 2014. "Android's App Permissions Were Just Simplified — Now They're Much Less Secure." www.howtogeek.com/190863/androids-app-permissions-were-just-simplified-now-theyre-much-less-secure/.

Hu, Wen, and Yanli Zhao. 2014. "Analysis on Process Code Schedule of Android Dalvik Virtual Machine." *International Journal of Hybrid* IDC, 2020. www.idc.com/promo/smartphone-market-share/os.

IMPACT. 2018. "Sand Droid Is An Automatic Android Application Analysis System." *In Information Marketplace for Policy and Analysis of Cyber-Risk &Trust (IMPACT)*, commons.stage.datacite.org/doi.org/10.23721/100/17388.

Jing, Yiming, Ziming Zhao, Gail-Joon Ahn, and Hongxin Hu. 2014. "Morpheus: Automatically Generating Heuristics to Detect Android Emulators." In *Proceedings of the 30th Annual Computer Security Applications Conference*, 216–225. ACM, New York, NY, USA.

Karim, Md. Yasser, Huzefa Kagdi, and Massimiliano Di Penta. 2016. "Mining Android Apps to Recommend Permissions." In *2016 IEEE 23rd International Conference on Software Analysis, Evolution, and Reengineering (SANER)*, 427–37. Institute of Electrical and Electronics Engineers (IEEE). doi.org/10.1109/saner.2016.74.

Kaspersky, 2014. "Number of the Week: List of Malicious Android Apps Hits 10 Million." www.kaspersky.com/about/press-releases/2014_number-of-the-week-list-of-malicious-android-apps-hits-10-million.

Khan, Shahnawaz, and Thirunavukkarasu Kannapiran. 2019. "Indexing issues in spatial big data management." In *International Conference on Advances in Engineering Science Management & Technology (ICAESMT)-2019*, Uttaranchal University, Dehradun, India.

Lockheimer, Hiroshi. 2012. Android and Security. googlemobile.blogspot.com/2012/02/android-and-security.html

Maier, Dominik, Mykola Protsenko, and Tilo Müller. 2015. "A Game of Droid and Mouse: The Threat of Split-Personality Malware on Android." *Computers and Security* 54 (October): 2–15. doi.org/10.1016/j.cose.2015.05.001.

Material.io. 2017. "Metrics & Keylines." material.io/archive/guidelines/layout/metrics-keylines.html#metrics-keylines-keylines-spacing.

Oberheide, J., and C. Miller. 2012. "Dissecting the Android Bouncer." jon.oberheide.org/files/summercon12-bouncer.pdf.

Ongtang, Machigar, Stephen McLaughlin, William Enck, and Patrick McDaniel. 2012. "Semantically Rich Application-Centric Security in Android." *Security and Communication Networks* 5 (6): 658–73. doi.org/10.1002/sec.360.

OpensourceForU. 2013. "What A Native Developer Should Know About Android Security?" www.opensourceforu.com/2013/09/what-a-native-developer-should-know-about-android-security/.

Petsas, Thanasis, Giannis Voyatzis, Elias Athanasopoulos, Michalis Polychronakis, and Sotiris Ioannidis. 2014. "Rage Against the Virtual Machine: Hindering Dynamic Analysis of Android Malware." In *Proceedings of the Seventh European Workshop on System Security - EuroSec '14*, 1–6. ACM Press, New York, New York, USA.

Rotterdamopendata. 2020. "Rotterdam Open Data." rotterdamopendata.nl/dataset.

Schwartz, Edward J., Thanassis Avgerinos, and David Brumley. 2010. "All You Ever Wanted to Know about Dynamic Taint Analysis and Forward Symbolic Execution (but Might Have Been Afraid to Ask)." In *Proceedings - IEEE Symposium on Security and Privacy*, 317–31. doi.org/10.1109/SP.2010.26.

Seppala, Timothy. 2015. "Google Won't Force Android Encryption by Default." www.engadget.com/2015-03-02-android-lollipop-automatic-encryption.html.

Singh, Shirish Kumar, Bharavi Mishra, and Poonam Gera. 2015. "A Privacy Enhanced Security Framework for Android Users." In *2015 5th International Conference on IT Convergence and Security, ICITCS 2015 - Proceedings*. Institute of Electrical and Electronics Engineers Inc. doi.org/10.1109/ICITCS.2015.7292926.

Statcounter. 2021. "Mobile & Tablet Android Version Market Share Worldwide." gs.statcounter.com/android-version-market-share/mobile-tablet/worldwide.

Techrepublic. 2012. "TaintDroid: Warns about Android Apps Leaking Sensitive Data." www.techrepublic.com/blog/it-security/taintdroid-warns-about-android-apps-leaking-sensitive-data/.

Tracedroid. 2012. Tracedroid-Dynamic Android App available on Github. github.com/ligi/tracedroid.

Trend Micro 2012. "A Look at Google Bouncer." www.trendmicro.com/en_us/research/12/g/a-look-at-google-bouncer.html.

Vidas, Timothy, and Nicolas Christin. 2014. "Evading Android Runtime Analysis via Sandbox Detection." In *ASIA CCS 2014 - Proceedings of the 9th ACM Symposium on Information, Computer and Communications Security*, 447–58. Association for Computing Machinery, Inc, New York, NY, USA. doi.org/10.1145/2590296.2590325.

Weichselbaum, Lukas, Matthias Neugschwandtner, Martina Lindorfer, Yanick Fratantonio, Victor van der Veen, and Christian Platzer. 2014. "{ANDRUBIS}: Android Malware Under the Magnifying Glass," no. February.

xda-developers.com 2014. "Play Store Permissions Change Opens Door to Rogue Apps." 2014. www.xda-developers.com/play-store-permissions-change-opens-door-to-rogue-apps/.

Yuksel, Asim S, Abdul H Zaim, and Muhammed A Aydin. 2014. "A Comprehensive Analysis of Android Security and Proposed Solutions." *Computer Network and Information Security* 12: 9–20.

Zhou, Yajin, Xinwen Zhang, Xuxian Jiang, and Vincent W. Freeh. 2011. "Taming Information-Stealing Smartphone Applications (on Android)." In *Lecture Notes in Computer Science (Including Subseries Lecture Notes in Artificial Intelligence and Lecture Notes in Bioinformatics)*, 6740 LNCS: 93–107. Springer Verlag, Berlin, Heidelberg. doi.org/10.1007/978-3-642-21599-5_7.

4 Machine and Deep Learning Techniques in IoT and Cloud

T. Genish and S. Vijayalakshmi

CONTENTS

DOI: 10.1201/9781003138037-4

4.1 INTRODUCTION

Rapid advancements in the software industry and communication technologies supported the development of sensor devices connected to the internet. These collect data from the physical world and it is used for decision support systems, data retrieval systems, and the like. For the last decade, complex tasks have started here, accomplished by the specific applications, and based on components such as mobile devices, actuators and sensors. Nowadays, every electronic device gets connected to the internet and this has led to the development of new era of the Internet of Things (IoT). The data collected from the IoT is analyzed into meaningful information. It paves the way for new techniques such as Artificial Intelligence (AI), Deep Learning (DL), and Machine Learning (ML).

A network attack is the main challenge when IoT is implemented. These attacks are categorized into two types: active and passive. In active attacks, the attacker gains unauthorized access as well as the ability to modify data in terms of deleting, encrypting, or harming it. As far as passive attacks are concerned, an attacker gains access to reach network monitors or steals sensitive information. The information has not been changed. K-Nearest Neighbor (KNN), Principal Component Analysis (PCA), Support Vector Machines (SVM) are popular ML and DL based methods used to avoid network intrusion and network anomalies.

In recent years, the healthcare system has an inevitable information technology network to deliver smart and accurate information in order to make diagnostics and treatment faster [Greco et al. 2020]. This system manages data in order to monitor the health conditions of patients and to provide medical automation in various environments such as hospitals, admin offices, homes, and so forth. It gradually reduces the cost incurred for a doctor's visit [Akmandor and Jha 2017]. As the technology has developed, medical sensors are embedded with powerful hardware devices to create a ubiquitous healthcare network called, the "Internet of Medical Things (IoMT)". IoMT changes the approach to healthcare such that the numbers of healthcare devices and wearable technologies implemented by IoT is expected to attain 162 million by end of year 2020 [Akmandor and Jha 2017].

Data obtained from embedded sensors, wearable and mobile devices helps to harvest information describing the everyday information about a user. The information is then gathered and processed to observe unfavorable conditions, using latest approaches such as AI, ML and DL. The cloud, which is used traditionally for big data analysis, gives a better reliability and performance to support IoT applications [He, Yan and Xu 2014]. In IoMT, the patients are end users and if they have time-sensitive requirements, a high degree of robustness is required. In this situation, disconnection from the network or latency has a negative impact and might have fatal consequences [Tang et al. 2019]. Fog/Edge computing deals with remote monitoring solutions in a healthcare system. The fog nodes relay information to a local server which gathers

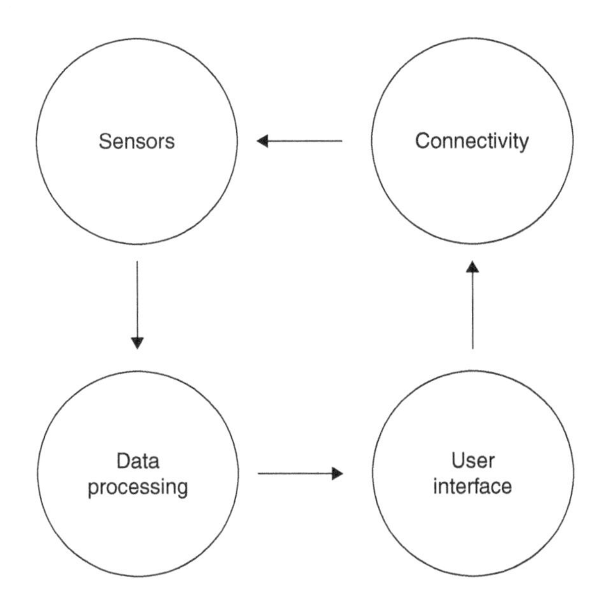

FIGURE 4.1 Components of IoT System.

and processes the data to give a quick respond [Satyanarayanan 2017]. Advanced technology in IoT which utilizes both software platforms and network architecture domains leads to the advent of smarter solutions. These solutions address different levels of healthcare such as pediatric and elderly care, disease monitoring, epidemic disease supervision and fitness management. Multi-layered architectures are involved in finding IoMT related solutions as shown in Figure 4.1.

In edge computing, devices like smartwatches, smartphones and embedded systems perform pre-processing steps and low-level operations on data gathered using Wireless Body Sensor Networks (WBSN). At the fog level, servers gather data from the sensor devices to support local processes and/or storage. The high-level computational tasks and data storage are performed at cloud level. Data required by the IoT is for the representation of useful services/interfaces to users or, for improving the IoT framework. In this view, IoT systems should be large enough to access data in raw format from various resources and process or analyze it, converting it into knowledge. Even though the IoT is the source of new data from different resources, data science also provides a remarkable contribution to make the IoT stronger and more robust.

The population explosion creates many challenges for healthcare and leads to scarcity of medical resources. Hence, IoT based healthcare services came into existence to give sufficient user level experience at very low cost, thus improving the quality of life. The important function of the IoT is to offer connectivity between medical resources and efficient, consistent, and smart healthcare services to sick and/or aged people who are affected by chronic diseases. The IoT is composed of smart functioning sensors, remote servers, and networks to bring the smart healthcare environment to medical services.

A collection of applications and tools to get computer networks linked in healthcare systems is called, the Internet of Medical Things (IoMT). Various tools are used to automatically monitor the health status of patients. The tools also manage real time information and reduce overall costs.

In the field of agriculture, the most important part is image analysis. It is used to identify or to classify anomaly detection in an image that has been input. ML and AI techniques influence smart farming in terms of automated agricultural growth, precision farming, environmental forecasting, and so on. The volume of data produced by sensors is high in smart-farming applications and is categorized into: structured, semi-structured, and unstructured data.

The information revolution has made it important to manage large amounts of data rapidly. The digital environment creates huge volumes of diverse and volatile data as all parts of the world is highly populated with IoT devices and IoT sensors [Alahakoon, Nawaratne and Xu 2020]. Nowadays, the applications of ML are high in the development of the smart city. The smart city model has been presented using ML techniques [Habiazdeh et al. 2019]. The smart city infrastructure is portioned into application, communication, sensing, data planes, and security. In this model, two components are investigated, (i) implementation of centralized and distributed infrastructure and (ii) application of modes such as machine learning, deep learning, data analytics, and data visualization to implement smart city software.

DL and ML algorithms are used in smart farming for various applications. DL gives effective solutions in handling complex problems such as picture recognition, detecting objects, image classification and extraction [Mohammadi et al. 2018]. The concept of DL is also applied to multiple functions such as automatic farming forecasts and detection.

IoT technology is applied in various real-life applications. Among them, implementation in the field of agriculture gives benefits to people living all over the world. Guo et al. analyzed the smart city using Mobile Crowd Sensing and Computing (MCSC). In this model, aspects such as crowd-sensing, and the integration of machine intelligence and human intelligence are discussed. The method forms a collaborative operation between human beings and machines where humans supervise the sensing and computational power of machines, with machines processing the raw data and performing decision-making processes.

A fully automatic smart irrigation system has been developed by Shekhar at al. In this model, sensors are used to capture data about soil temperature and moisture and the KNN classification method is used to get sensor data for the prediction of soil water irrigation. An IoT based hydroponic framework has been developed by [Mehra et al. 2018]. This method offers control in the hydroponic environment with various input parameters such as temperature, lighting, moisture level, and pH, without human intervention. A framework for the classification of a leaf and the identification of plant diseases has been developed with the help of convolution networks [Sladojevic et al. 2016]. The proposed method can identify 13 different types of plant diseases. To create a database, 3000 pictures have been collected from internet sources

Smart farming to predict suitable yields has been designed using DL and Wireless Sensor Networks (WSN) [Miles 2020]. This method improves the field irrigation framework and continuously monitors the field. Soil parameters such as moisture, temperature, and other necessary data, are collected for the purpose of field surveillance. WSN retrieves information from the sensors and analysis is completed through uploading cloud information.

Recommendations have been made for the implementation of mobile phone-oriented platforms that collect data from various sources such as the internet and from sensors to implement ML techniques for the prediction of future outcome [Pejovic 2015]. One recent study [Siow et al. 2018] has revealed the development of IoT and data analytics. In this model, parameters such as domain, objectives, resources and frameworks are investigated.

To address transportation problems, the Support Vector Machine (SVR) method is used for the prediction of traffic flow in typical and atypical situations [Castro-Neto et al. 2009]. A DL approach is implemented to extract the characteristics of the problem from the data. The method is applied with SVR since SVM consumes high memory resources. In a video sequence, tracking aims to locate a target and to give the location of the first frame in which the object appears. This is used in surveillance systems. It is an important task to track any suspects or suspect vehicles automatically for the application of safety monitoring and urban flow management. A robust tracking algorithm is presented in [Li et al. 2016] which uses feature representation of a target object. The end-to-end object tracking mechanism is applied to map raw sensor input to object tracks.

A graph-based technique is proposed [Rathore et al. 2015] to produce a smart transportation system. Information about overall traffic is obtained using sensors and the vehicular network is used to gather data about speed and location of the vehicle. This IoT based model generates huge volumes of data, called big data. To manage the processing of big data, a tool, named Giraph is implemented with parallel processing servers. To represent a real city's traffic, vehicular datasets are collected from various resources and used for analysis and evaluation. The performance of this graph-based technique is calculated in terms of efficiency, with particular reference to system throughput and processing time.

IoT has been identified as an effective solution for medical emergency and other healthcare systems. It monitors patients who are suffering from serious health disorders such as Parkinson's disease and diabetes. Emergency healthcare is also recognized but has not got a wide research base, yet. An automated healthcare monitoring system should provide not only fast access to data but also reliable data for accurate prediction of health problems by healthcare service providers. For this challenge, an e-healthcare system has been developed by using DL techniques [Jeyaraj and Nadar 2019]. The method is prototyped as an electronics component that uses intelligent sensors for signal measurement. To measure the performance of this system, physiological signal prediction accuracy is calculated.

A system to monitor the blood-glucose level in diabetic patients was developed [Chang et al. 2016]. This method requires diabetic patients to take blood-glucose readings manually at intervals. Then, two types of blood-glucose abnormalities are considered. The first category is abnormal blood glucose level and the other one is missed blood-glucose readings. An abnormality condition of a patient is then analyzed by the system which then sends notifications to the patient, caregivers, family members, and healthcare providers such as physicians. The system is proved to be an effective one in monitoring the health conditions of diabetic patients.

A method was developed to detect heart attacks through a custom antenna [Wolgast et al. 2016] using an Electrocardiogram (ECG) sensor. An ECG sensor collects information about heart activities that is then processed by a microcontroller. The collected information is transmitted to users' smartphones through Bluetooth where the ECG data is processed and presented in a user application. This method needs to be upgraded to measure the respiratory rate to further help in the prediction of heart attacks. Another method was proposed to monitor the activity and health of a human being that uses wearable, environmental and vision-based sensors [Zhu 2015]. This method allows older and chronically ill people to live in their homes comfortably and monitors their health. If any health issues arise, the intervention of doctors and caretakers is initiated. Future enhancements are in progress to this method to implement ML techniques to make decisions about patients' health.

4.2 BASIC CONCEPTS OF IOT

4.2.1 Working Mechanism of IoT

The IoT is a process of creating a network of physical objects known as things. These things have been embedded with components such as software, electronics, network, and sensors. The embedded systems allow the objects to collect and exchange data. The IoT creates everything virtually, which makes the lives of people smarter. The point about the IoT is that it is not only physical devices but also a person who has had a chip inserted that monitors the diabetes level in his body or an animal that has had tracking devices inserted. The functioning of the whole IoT system is based on four primary components, given below. The components of the IoT system are shown in Figure 4.1.

- Sensors/Devices
- Connectivity
- Data Processing
- The User Interface

4.2.1.1 Sensors and Devices

The first component of an IoT system is a sensor that helps in the collection of precise, fine data from the physical environment. The collected data may have complexity ranging from simple location identification to extended videos. The sensors for temperature, light, and other electromagnetic spectrum measurements, are used based on requirements. For instance, wearable devices and smartphones are embedded with sensors, particularly gyroscope and accelerometer. Data collected from the sensors is used in various applications such as the recognition of human activities, stability in medicine, availability of parking spaces and so on. The sensors have been chosen, based on required parameters and precision. The parameters may be accuracy and reliability of collected data.

4.2.1.2 Connectivity

Here, the collected data is required to be sent to a cloud infrastructure. The sensors connect with the cloud over transport and communications media such as satellite networks, Bluetooth devices, cellular networks, Local Area Networks (LAN), Wide Area Networks (WAN), and so forth. The appropriate network infrastructure may be chosen based on the intended use. There is a trade-off between power consumption, bandwidth, and range of specifications available. It is important to choose the appropriate mode of connectivity wisely for the IoT system.

4.2.1.3 Data Processing

The data processing component of IoT follows three steps. These are: acquiring input, processing the input data and generating output for each input.

- Input: is obviously needed for any data processing. The collected data may be of any form such as: text, numerical values, images, QR codes and sometimes even audio and video. Before processing the data, all of the input readings must be converted into a machine understandable format.
- Process: is next to the input phase, where the actual processing of input data occurs. To convert raw input data into meaningful information, multiple techniques such as sorting, classification, clustering, and so forth, are applied.
- Output: provides a readable form for information. Although meaningful information is generated at the processing stage, it should be deciphered into human readable form prior to the output stage. The rendered output may be in the form of text, table, graph, numbers, audio, or video formats, for example. Subsequently, the output is stored as data for processing or stored for future reference. Storing the output data is an important process since the comparison between current data and historical information helpfully inform the better functioning of a system.

4.2.1.4 The User Interface

The received output will need to be shared with an end-user in such a way as to activate alarms in their electronic gadgets, send the notifications to their email or send text messages to their mobile phones. The end-user might also need an interface in order to actively check their IoT systems. For instance, if a user had installed a camera in his home, he will need to access the video footage and any associated information from a web server. The user might be able to act, based on the application of an IoT system and dependent on its complexity. For example, if a user desired to detect any changes in the temperature of an air-conditioner, then this could be followed up by adjustment of that temperature using a mobile phone.

4.2.2 Importance of IoT

All over the world, people can live happier, work smarter, and achieve full control over their family and business lives because of the advent of IoT. IoT not only offers smart devices for home automation but also provides optimized solutions for business

[Zhao et al. 2020]. IoT provides consistent results to most applications where human resources are enhanced by machines. It allows those engaged in businesses to effectively model the operations of their various systems.

IoT helps industry to automate processes and minimize labor costs. It also reduces manual work, enhances service delivery and reduces the cost of manufacturing and delivering goods. It also offers transparency in customer transactions. IoT is an inevitable technology use in our everyday lives and increasingly, the business world is coming to realize its potential.

4.2.3 Applications of IoT

The IoT exists in all areas of family life, in various organizations, businesses, government projects and so on. Here, some of the important aspects of IoT for various applications are described.

4.2.3.1 Smart Homes

The smart home is one of the very interesting applications of IoT that many people in the world would like to enjoy. This application brings us to a new level in terms of convenience and security. Even though the IoT provides many advantages in its application in the smart home, one of its best is in entertainment and other utility systems. For example, an IoT system embedded with the electricity meter, a meter to measure consumption of water, a set-top box that gives an option to record shows, surveillance systems, security systems, lighting systems, and so on. As the technology of IoT emerges, devices will be smarter, thus enabling high end home security.

4.2.3.2 Transportation

In every layer of transportation, IoT brings improvements in communications, control and distribution. People in private and government organizations have started applying IoT technology to transportation problems. The transportation domain is the second largest one to invest in Industrial IoT (IIoT). Advancements in mobile and communication technologies have made IoT smarter. Electronic gadgets play a leading role as they monitor operations for increased safety and efficiency, as shown in Figure 4.2. A popular IoT solution company, Biz4Intellia has listed five ways in which IoT can support the transport sector.

1. Fleet Management: It is an important management goal to make an organization more cost effective. This technology is growing gradually with developments in terms of maintenance cost, operational efficiency, fuel consumption and so on. In addition to this, fleet management offers some key features such as GPS tracking, customized dashboards, geo-fencing, and so forth.
2. Public Transport Management: Real-time tracking of a vehicle is achieved by IoT, the tracked data is transferred to a control engineer or to a centralized system and then forwarded to internet connected devices.

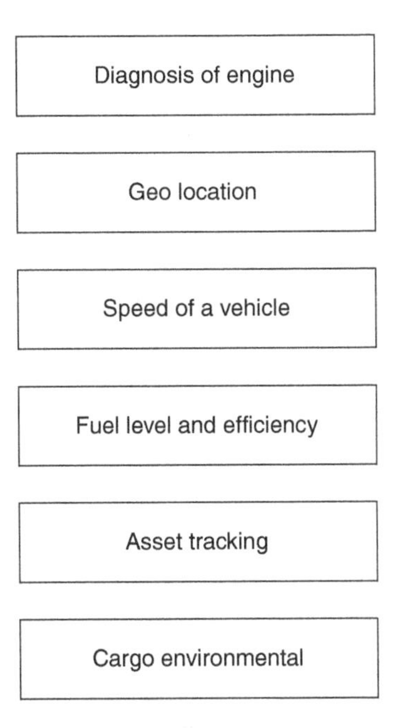

FIGURE 4.2 IoT mechanism in transportation.

3. Smart Inventory Management: This is used to provide information about warehousing, production, and distribution on a real-time basis. Collated information is then utilized to reduce the consumption of goods on the inventory and to improve predictive maintenance. Accurate data collected from IoT sensors facilitates a robust inventory management system.
4. Optimal Asset Utilization: This is applied to manage physical assets: their location, status, and so on. Information regarding the exact latitude and longitude of an asset are also obtained. Information is provided regarding the threshold properties of a device, using sensors.
5. Geo-Fencing: This is a development, making use of GPS, that finds the current location of an asset / device with map references of its geographical area. With this technology, a driver is sent alerts when he / she has deviated from an actual route and hence accidents are avoided.

4.2.3.3 Agriculture

The field of agriculture has seen various transformations in terms of technology. The term 'Smart Agriculture' has developed and it refers to the application of IoT in farming.

IoT helps in monitoring and managing micro climatic conditions and thus increases crop productivity in indoor planting. As far as outdoor planting is concerned, IoT helps in sensing the moisture levels of soil and of nutrients. It also assists farmers in taking farm related decisions, by informing them of the quantity of fertilizers and pesticides needed to achieve optimal efficacy. The role of IoT in smart farming is shown in Figure 4.3.

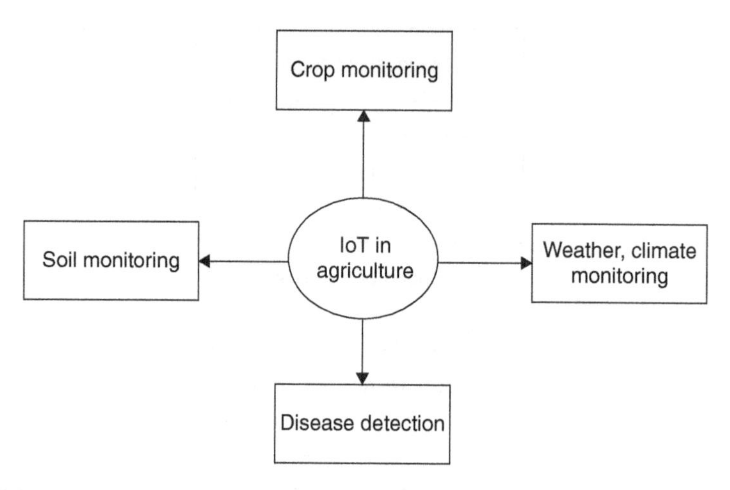

FIGURE 4.3 Applications of IoT in Smart Agriculture.

4.2.3.4 Healthcare

The application of IoT in the healthcare field is moving slowly. As per the research conducted in 2020, 40% of IoT based devices have been used in the healthcare industry. IoT devices monitor the patients on a real-time basis and save lives during medical emergencies such as heart attacks, respiratory problems, diabetes, and so forth. The devices connected to the IoT platform collect and transfer health related information such as blood pressure, weight, oxygen and sugar levels in the blood, and ECGs are illustrated in Figure 4.4. The collected data is stored in cloud architecture and shared with the physician, consultant, or any authorized person. Patient care is automated with the help of next generation healthcare mobility technologies. By implementing IoT, medical errors are reduced significantly.

4.2.4 Considerations on the Use of IoT

4.2.4.1 Advantages

IoT sensors make our everyday life simpler and smarter in the fields of business, healthcare, agriculture, and so forth. Nowadays, IoT technology is popular because of the advantages that it brings. Some of the most important merits of IoT are listed here.

Collecting information: Real-time data can be accessed easily from anywhere in the world. Users can work on the data even though they are not physically present.

Saving time: Smart devices assist in retaining large numbers of records, whereas, creating all of the records manually, would be time consuming and costly in terms of labor.

Easy communication: In IoT technology, communication devices offer transparent and accurate data transmission. Machine to machine communication is performed efficiently and results can be obtained more quickly.

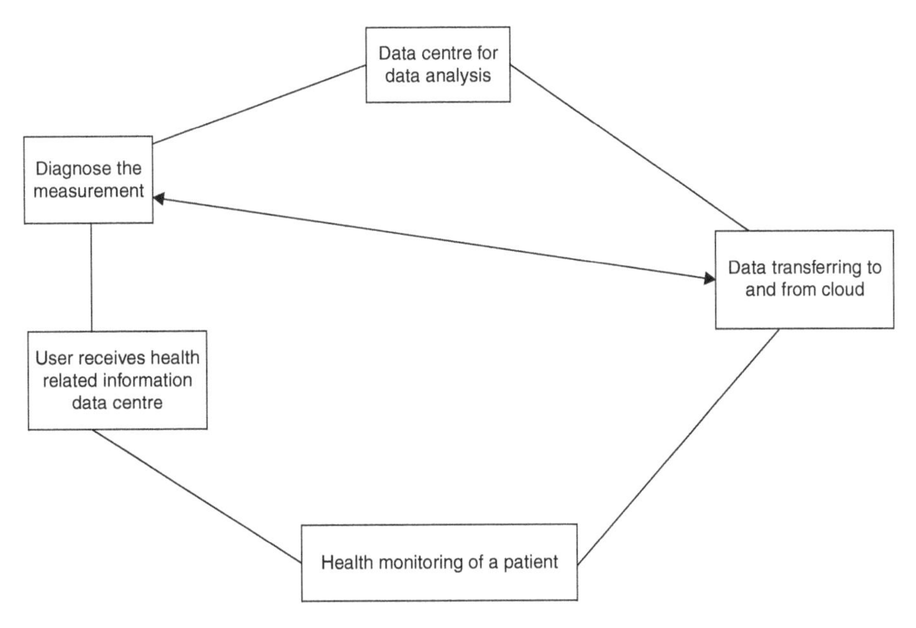

FIGURE 4.4 IoT processing in Healthcare System.

Automation: As the tasks are automated, industry is offered a high quality of service with reduced human intervention.

4.2.4.2 Disadvantages

Even though the services offered by IoT are smarter and accurate, it has some disadvantages or limitations as listed below.

Security threats: Since the IoT mechanism is interconnected with various networks, it creates a pathway for hackers to execute network attacks.

Compatibility: A compatibility problem occurs in the IoT as devices from different manufacturers are connected. There are no international standards for compatibility of equipment.

Complexity: The IoT is a collection of diverse and complex networks. Malfunctions or failures in software or hardware components can lead to severe consequences. Power failures may also cause many problems.

Technology controlling human life: As human lives are under the control of technology we will come to depend completely on it. Even for the little things that we do, we may get addicted to the devices or machines.

4.3 CLOUD COMPUTING

Cloud computing is a technology which both large and small sized organizations use to store their data in the cloud environment and access it anywhere at any time.

Cloud computing provides services such as storage, software, databases, and other platforms that can be accessed through the internet.

4.3.1 Cloud Computing – Architecture

The architecture of cloud computing is one which has embedded components to offer services to users in terms of storage, the sharing of data, maintenance, and flexibility. Cloud applications and services such as Amazon, or Netflix are accessed by the internet or through a virtual network. Organizations have gradually moved into cloud technology as it provides significant storage and it combines both service-oriented and event driven approaches. Cloud computing is classified into two categories, namely front end and back end. The front end is for the client and consists of interfaces and applications needed to access the cloud computing platform. The back-end category is for the service provider and is comprised of the resources needed to offer cloud computing services. Resources may be of storage, deployment models, decision making strategies, security mechanisms, and so on.

4.3.2 Cloud Computing – Services

There important cloud computing services are as follows:

1. Infrastructure as a Service (IaaS): IaaS offers accessing to storage, servers, connections, and so forth. External cloud providers offer virtualized infrastructure for business management. For example, IaaS providers are Microsoft Azure, Google Cloud Platform, Amazon EC2, and the like.
2. Platform as a Service (Paas): PaaS is built on IaaS. In this environment, the cloud vendor delivers resources such as hardware and software infrastructure components (such as operating systems and middleware). PaaS allows cloud users who are accessed through a webpage to install data sets, development tools, and so on.
3. Software as a Service (Saas): SaaS consists of both IaaS and PaaS. The software suite is offered by a cloud service provider on a pay per use basis. Popular examples of SaaS include Microsoft Office 360, Adobe Creative, Zoho, Salesforce, Oracle CRM, and others.

4.3.3 Cloud Computing – Types

Cloud computing is categorized into public cloud, private cloud, hybrid cloud and community cloud.

The public cloud is managed by a third party that provides services to the public over the internet. It offers solutions to minimize the cost of an IT infrastructure. It helps small enterprises to start a business without much investment. A public cloud needs to serve multiple users and the user may require a virtual computing platform which is separated and isolated from other users.

In a private cloud environment, the users are provided with dynamic provisioning of resources. In this type of cloud, the usage of the cloud is charged for on a proportional billing basis instead of the pay as you use model used in the public cloud. Major advantages of using private clouds are i. that customer information is protected, ii. that there is Infrastructure ensuring SLAs, and iii. that there is compliance with standard procedures.

A hybrid cloud is a combination of a public and a private cloud system, distributed heterogeneously. The hybrid cloud is otherwise referred to as the heterogeneous cloud. It is unable to scale on demand to addresses peak loads efficiently. This is considered to be a major drawback of a hybrid cloud but, it does have the benefits of both public and private clouds.

The community cloud is a distributed system that integrates different cloud environments to address the requirements of an industry, community, or business. Communities that use these clouds are media, healthcare, scientific research, and energy industries.

4.4 IOT AND CLOUD COMPUTING

The need for the IoT in the world is growing rapidly and the various IoT applications are used in many business models. Future technology is strongly based on the IoT and cloud computing where each model provides another platform. The advantages of IoT and cloud computing have been discussed below.

4.4.1 Benefits of Using IoT and the Cloud

Scalability for device data: The most important advantage of the IoT in the cloud is scalability. To meet the needs of data analysis, cloud-based solutions can be scaled vertically and horizontally. For instance, the server capacity can be increased to host more applications when necessary. Additionally, the IoT cloud platform gives more flexibility to deal with storage limitations.

Data Mobility: As data is stored and processed in a cloud environment, it is easy to access data from any location. Data is not bounded by any infrastructure or network components. The IoT cloud platform provides tools for companies to manage, provision and update devices and sensors.

Time to Market: The IoT cloud platform gives solutions in a shorter time and gives support to achieve a low implementation cost. From a business point of view, it is profitable to analyze and arrange upgrades to a company's network structure [Kahn et al. 2020].

Security: This is a major challenge to all the organizations across the world. In the case of any clashes in the infrastructure, the IoT cloud will take over responsibility. Some companies may feel uncomfortable about giving control to a third party. Hence, it is in hands of the organizations to maintain good security policies and practices.

Cost Effectiveness: Issues such as cost incurred for the maintenance of hardware and IT personnel are always of concern. A payment scheme based on usage will give encouragement to organizations to move towards IoT cloud scenarios.

4.4.2 Importance of Cloud and IoT

Technology continues its evolution day by day and the things around us become connected. Due to agility and scalability, IoT cloud integration creates a revolution in industry. IoT systems collect data stored in the cloud environment and perform intelligent decision making. The cloud is an important vehicle for data aggregation and for drawing insights from the data. IoT and cloud technologies are complementary to each other. Both the technologies have been utilized to increase the efficiency. While IoT generates massive data from sensors, the cloud creates a way of handling this huge data. IoT and cloud have built integration for the solutions of data storage and data access. The cloud offers solutions and provides scalability while accessing remote data through dedicated internet channels. The cloud make is possible to resolve IoT driven business needs. In many ways, it has been undoubtedly proven that the cloud environment is essential for the success of IoT.

4.4.3 How Can the IoT and the Cloud Be Expanded?

Data produced for cloud technology is increasing day by day. Big data is the new approach to processing and analyzing data [Kahn et al. 2019b]. The benefits of applying big data concepts to the IoT are that they provide scalable and reliable solutions for business organizations. Instead of generating raw data and sending it to the cloud for analysis, a new approach has appeared for data storage and data analytics, called **fog and edge computing**. It is a system of moving data to local gateways such as routers or switches. This edge device then performs the needed processes and returns decisions to IoT devices. It is termed as fog/edge computing.

4.5 MACHINE LEARNING AND DEEP LEARNING IN IOT AND CLOUD

Daily, an ever-increasing number of IoT devices are connected and communicate with each other and produce large amounts of data every day. IoT devices are also programmed in many applications to perform tasks either based on predefined conditions or based on feedback from collected data. In this situation, human intervention is needed to analyze the data and extract knowledge from the raw data to create smart applications. They must be able to take decisions on the basis of contexts called Cognitive IoT (CIoT).

ML is part of the AI system. The computer system carries out tasks such as clustering, classification, recognition, prediction and so on with the help of ML techniques. The system is trained to complete a learning process using statistical models and analyzing sample data. The sample data is characterized by certain features and a ML technique finds any correlation of features and output values called labels [Al-Dweik et al. 2017]. ML algorithms can be categorized into four types.

a) Supervised Learning: This deals with a problem that involves regression, like forecasting weather, or the prediction of population growth by applying techniques such as Random Forest or linear regression. In addition to this, supervised learning gives a solution to the problems occurring during classification, such as speech and digit recognition and diagnostics. These problems are addressed by the implementation of Support Vector Machines (SVM). The training phase and the testing phase are the two phases involved in supervised learning.

b) Unsupervised Learning: This addresses the problems involved in dimensionality reduction used for feature extraction and big data visualization. It also deals with problems such as customer segmentation, recommendation systems and targeted marketing. Unsupervised learning algorithms identify patterns when testing and clustering data or they predict feature values.
c) Semi-supervised Learning: It combines the characteristics of supervised and unsupervised learning methods. The working principle of semi-supervised learning methods resembles unsupervised learning but with the bonus of labeled data.
d) Reinforcement Learning: In this technique, the algorithm tries to predict output values to the problem, consisting of a set of tuning parameters. Then, the computed output becomes an input value and a new output is computed until an optimized output is found. Algorithms for reinforcement learning are applied for robot navigation, AI based gaming and real time decisions.

4.5.1 Overview of Deep Learning

DL is the part of Artificial Neural Networks (ANN) that mimics the functions of the human brain. Applications of DL include natural language processing, driverless cars, natural language processing and others. In the DL model, computers learn ideas from speech, text, and pictures and apply classifications. DL models are trained using huge sets of data which are labeled with neural networks architecture. Most of the DL models utilize the architecture of neural networks. Thus, DL models are known to be Deep Neural Networks (DNN). The DL training method is represented as "Deep" since Artificial Neural Networks (ANN) covers many levels as time goes by. The DL approach uses information processing algorithms and performs the process of thinking [Liakos et al. 2018]. DL and ML approaches are used to improve analytics and learning in the IoT.

4.5.2 K-Nearest Neighbor (KNN) Classification Model

This is a non-parametric and supervised learning paradigm where the data is classified into a given category using a training set. In KNN, a prediction is made of a new instance (x) in searching the entire training set for the K most similar cases (neighbors) and summarizes the output variables for that K cases. This value is a modal class value in the classification. The purpose of modal value is to use the database where the data points are divided into multiple classes for the classification of the new sample point. Some of the advantages of the KNN classifier are firstly that it is easy to implement; secondly, it offers fast computation for small training data sets and thirdly, it doesn't need prior knowledge about the structure of the data. The noted limitations of KNN are firstly, that is consumes more space if the training set is large and secondly, that the distance is computed between training data and test data and thus the testing takes a lot of time.

4.5.3 Machine Learning and Deep Learning in Smart Transportation

Smart transportation is an important part of research since it occupies increasing amounts of space in our day to day lives. Further, many difficulties have been addressed to establish and maintain the modern smart city. These problems are addressed by machine and DL techniques with the implementation of the IoT. The categories of smart transportation include route optimization, light, accident detection/prevention, parking and infrastructure.

The transportation field has been introduced to the latest IoT technologies and this has led to the invention of the Intelligent Transportation System (ITS). The field of ITS has received more attention amongst researchers since there are huge opportunities for future development. A most important and significant area of interest in smart transportation is route navigation (or route optimization). By using data from the mobile phones of users or with the units placed in particular locations on the road, ITS provides data about traffic congestion and gives optimized routing options. It helps in minimizing traveling time and thus reducing car emissions and energy consumption. IoT devices are also used to establish smart parking systems. With the help of cameras or other sensors such as magnetic field or IR sensors, researchers have developed a smart parking system which allows maximization of the availability of parking space and minimization of the searching time. A new system is proposed to find anomalies in the surface of roads. This system is based on input data received from sensors attached to cars or the drivers' phones. IoT Machine to Machine (M2M) communication provides an opportunity to create vehicle to vehicle communication and vehicle social networks. It leads to the exchange of information with each other and creates possibility of the development of new applications.

Table 4.1 explains the importance of DL methods in smart transportation applications.

The main applications of DL in smart transportation are in the prediction of traffic flow, monitoring traffic and autonomous driving.

1. Prediction of Traffic Flow: The basic problem that the transportation model, and its management, faces is the prediction of traffic flow. Real-time and historical data are collected from various sensors including camera, crowd sourcing, and the like. A popular ML technique SVM (Support Vector Machine) is used to utilize such heterogeneous data efficiently.
2. Monitoring Traffic: Development of an automatic traffic monitoring/management system reduces manual operations by humans and warns drivers of dangerous situations. Traffic video analysis is an important part of traffic monitoring architecture. DL techniques are applied in video analysis from three perspectives: object detection, object tracking, and face recognition.

TABLE 4.1
Deep Learning Techniques Applied in Smart Transportation

Application	Model	Description
Traffic flow prediction	DBN, SAE	DBN in deep architecture is used for unsupervised feature learning. A stacked Auto Encoder (SAE) is utilized to learn generic traffic flow features
Short-term traffic prediction	LSTM	LSTM considers temporal spatial correlation in a stwo-dimensional network
Crowd flow prediction	CNN	Improves speed and accuracy in detecting objects
Real-time object detection	RPN, YOLO, SSD	Improves speed and accuracy in detecting objects
Object tracking	RNN	Tracking and classification of objects from static and moving platforms
Road accident detection	SDAE	Feature representation is learnt from spatio-temporal volumes
End to end learning for self-driving cars	CNN	Mapping raw pixels from front facing cameras to steering commands
Learning driving from video dataset	FCN-LSTM	Learn from crowd sourced vehicle action data

Object detection is applied in the situation where on-road vehicle detection, object detection, and pedestrian detection is required. Convolution neural networks are applied here to improve the accuracy of detecting the speed of an object. Object tracking is intended to identify and locate a target in the video sequences. Face recognition and detection methods based on neural networks are applied to monitor the activities of vehicle drivers and pedestrians.

3. Autonomous Driving: Mediated perception and behavior reflex are the two approaches in autonomous driving systems. In the mediated perception approach, the system computes high-dimensional world representation. It recognizes objects relevant to the driver such as traffic signs, lanes, cars, pedestrians, and so forth. The self-driving system is trained from the driving videos to learn from the driving videos and construct a direct map from the sensory input to a driving action. A learning-based approach is implemented to train a FCN-LSTM to predict continuous driving behaviors.

4.5.4 Machine Learning and Deep Learning in the Smart City

4.5.4.1 System Architecture and the Smart City

Applications for the smart city differ depending upon the requirements. A standard operational architecture of a smart city system consists of five planes: i. application, ii. sensing, iii. communication, iv. data and v. security planes.

4.5.4.1.1 Application Plane

This is an interface between the smart city and its users. The aim of an application is to reduce expenses by controlling misuse of resources, promoting task automation and improving safety and security standards using continuous and ubiquitous monitoring. This objective is achieved by creating links between the user and the data plane directly or indirectly. Many challenges are faced in terms of interoperability. These challenges come from the interactions between users and the data plane.

4.5.4.1.2 Sensing Plane

This incorporates various sensor devices and actuators that are used to calculate physical signals (environmental irradiation) and interaction with things (streetlights). Traditional WSNs are used for the implementation of the sensing plane. WSNs have some limitations and the availability of short battery life remains a major challenge. These limitations might be solved by using self-power-harvesting like solar and wind energy harvesting [Habibzadeh et al. 2017]. In the implementation of the sensing plane, the constant interactions between users and the environment implies that the plane should be flexible and expandable. In many situations, features like over the air firmware updates lead to additional privacy and security threats [Arias et al. 2015].

4.5.4.1.3 Communication Plane

The pre-processing and aggregating of data collected from the sensing plane are completed in the communication plane. The function of the communication plane is to support high-throughput, flexibility, low-latency, and highly secured communication. Data transmission is initiated at the request of on-node modules, to send the data to the cloud. Gateways are used by application specific routers in dedicated sensing mode while the smartphones of users can act as gateways in non-dedicated sensing. The architectures of single-hop and multiple-hop systems have differences in terms of congestion, latency, power consumption, and so on. Data aggregation is also a service provided by the communication plane which improves the quality of communication and ensures maximum life of battery-powered devices. Pre-processing techniques have also been applied for noise suppression and data extraction which helps data aggregation to remove repeated data in the earlier stages [Page et al. 2014].

4.5.4.1.4 Data Plane

The data plane is involved in converting acquired incoherent data into meaningful information. It offers two services: data processing and data storage, to other planes. The function of data processing involves processing hardware like CPUs and GPUs executing a wide range of algorithms. Data storage involves the storage, collection, and creation of databases for raw-data and meaningful information. Data storage is critical for various applications of smart city architecture because past trends in data provide a basis for authentic evaluation of newly collected data. Due to this, long-term data storage is associated with new and advanced DBMS tools. To provide functions such as data processing and data storage, the data plane integrates three techniques: data analytics, ML algorithms and data visualization techniques.

TABLE 4.2
Comparisons of Methods Implemented Using Cloud and Edge-Based C

Method	Architecture	Advantages	Disadvantages
Cloud Storage	Wireless Mesh-based [Shwe et al. 2016] SOS Framework [Dey et al. 2015] Data Replication and Reduction-Based [Yin et al. 2012]	Scalability, ubiquity, low cost, data centralization, Easy deployment, backup and recovery	Bandwidth based performance, single failure point, security issues, dependence on service providers, delays caused by distance.
Edge Storage	Fog server based [Luan et al. 2015] Cloudlet based [Stojmenovic et al. 2014] STACEE [Neumann et al. 2011] ACMES [Wu et al. 2017]	High scalability, low cost, real-time suitability, low latency, avoiding server traffic congestion	Lack of highly secured methods, data localization at edge

4.5.4.1.5 Security Plane

Even though there have been many breakthroughs in the IoT, it faces many security threats. The increase of cyberattacks in the IoT field have raised awareness about the importance of privacy and security issues [Cabaj and Mazurczyk 2016]. The security plane should give protection to all the components of the system uniformly, otherwise vulnerability may impact the entire system badly. Problems in ensuring security and privacy vary from one plane to another plane of the architecture of the smart city. Customized solutions are required to provide device-level security against software and hardware attacks. The main concerns of the smart city are spoofing, data leakage and Distributed Denial of Service (DDoS) attacks. Conventional cryptographic methods such as Elliptic Curve Cryptography (ECC), and the Advanced Encryption Standard (AES) provide good immunity against these attacks. In addition to this, many threats and vulnerabilities have come from characteristics of the IoT. The dynamic nature of smart city architecture and the mobility of devices makes the identification and authentication of participants more complicated. Table 4.2 gives a comparison between cloud and edge-based implementations and their merits and weaknesses in terms of cost, backup and security.

4.5.5 Machine Learning and Deep Learning in Agriculture

DL algorithms have been utilized in various different applications in agriculture such as fruit type, fruit counting, crop or plant classification, and so forth. DL techniques are also used to solve complex problems such as picture recognition, natural language processing, image classification, object detection, and image segmentation. The most used DL techniques used in smart-farming systems are discussed in this section.

Convolutional Neural-Networks (CNN): CNNs are a collection of artificial neural networks that are used for image analysis. A CNN is made up of neurons that have learnable weights and biases. It accepts 2D input data such as an image or voice and creates characteristics through a sequence of hidden layers. CNN are used in the agricultural sector for the detection of crop and plant leaf diseases.

Recurrent Neural Networks (RNN): These are networks of neuron-like nodes that are composed into layers. Every single node is related to a directed connection that has been utilized for an agricultural purpose. such as crop yield estimation, weather prediction, soil cover classification, soil moisture estimation, and so forth.

Generative Adversarial Networks (GAN): The GAN model is used to inspect and interpret using the training dataset. It is the framework which combines two neural networks such as generative and discriminative networks. These networks have joined together to produce high quality data.

Long-Short Term Memory (LSTM): Among various DL techniques, LSTM is the most common technique that can process single information points (such as images) and whole parts of data (such as voice or video). This model is suitable for classification and forecasting on time series data. From the smart-farming point of view, it is used for crop yield prediction, crop type classification and weather prediction. In addition to this, LSTM is applied for handwriting recognition and speech recognition. Previous works carried out in DL techniques in the applications of agriculture are given in Table 4.3.

To make progress in smart farming, the important agricultural issues addressed using the IoT is given in Table 4.3.

TABLE 4.3
Problems in Smart Farming

Problem	Description
Problems in irrigation	Proper water management is needed and water management systems should be used intelligently [Muangprathub et al. 2019]
Lack of soil knowledge	Structure of soil is changing everyday due to weather. So, the farmers are facing problems in finding the soil for crop production [Manikandan et al. 2020]
Problems in finding diseases in plant	Detection of plant diseases at the right time is necessary. Automatic detection of such diseases is required [International Atomic Energy Agency, "Agricultural Water Management."]
Logistics Management problem	Location-based sensor improves the efficiency of supply chain solutions. It gives transparency and client understanding [Razzak et al. 2018]
Detection of nutrient deficiencies	IoT devices may provide assistance to evaluate nutrients level in soil and plants [Sundmaeker et al. 2015]
Detection of nitrate level	Monitoring nitrate level in soil, water, fruits and vegetables [Alahi et al. 2018]

4.5.6 Machine Learning and Deep Learning in Healthcare

IoT architecture in the healthcare system consists of hardware and software components. The hardware comprises of temperature sensors, heart rate sensors, and blood pressure sensors. The processing of these sensors has various stages: collecting a sensor value, storing the data in the cloud, and data analysis to check abnormal conditions in patients. The software component consists of a software module/algorithm to interface and operate the hardware. Health conditions in the patient are monitored and analyzed by using ML algorithms. A classifier is a systematic approach to build classification models from input data. Machine and DL algorithms such as rule-based classifiers, Adaboost (short for adaptive learning), classifiers, support vector machines, and Naive Bayes classifiers, are commonly used techniques. Each method uses a learning algorithm to identify the model which best fits the bond of an attribute set and class label of input data. The aim of each learning algorithm is to create a model with better generalization capability.

ML algorithms achieve a higher level of accuracy in IoT based healthcare systems. In the stress detection model, the four categories of: data, feature extraction, classification and assessment are involved. For data, the ECG signals of vehicle drivers have been collected from the MIT-BIH database. The dataset is the part of Picard & Healey experimentation [Healey and Picard 2015]. The experimental data was collected from 17 drivers. The raw data consists of time, Electromyography (EMG), respiration, ECG, galvanic skin response (GSR) and intermittent heart rate (IHR). All these data have been collected from wearable sensors. After the datasets are collected, feature extraction is done using NetBeans Java. The stress level from ECG signals is grouped into 3 classes named, class 0, class 1, and class 2. Class 0 represents low-level stress; class 2 represents medium level stress and class 3 represents high level stress. ML algorithms such as SVM, Naive Bayes, logistic regression, ZeroR, IB 1 (1 nearest neighbor), IBK (k nearest neighbor), J48 (decision tree), random tree, and random forest are used for various data classification problems. Assessment is based on the time monitoring of automobile drivers to notify low or moderate levels of stress. If the stress rate is high, then the driver will be advised to concentrate on driving or take rest. The limitations of IoT based healthcare systems are given in Table 4.4.

4.6 CONCLUSION

The number of innovations that are being carried out day by day in the field of computer science is growing. Many smart technologies have been introduced to make our day-to-day life easier. One of the most notable technologies is IoT and AI (blended). This technology finds its application in almost all fields, for example: smart transportation, the smart city, agriculture, healthcare, and so on. In this chapter, the importance of cloud computing, the basics of the IoT, and various applications of IoT and AI (blended) techniques have been discussed. Merits and weaknesses of existing methods have also been discussed for better understanding.

TABLE 4.4
Limitations in Smart Healthcare Systems

Method	Contribution	Limitation
Limbs control and monitoring [Occhiuzzi et al. 2010]	Helps patients during night where tags are placed in their chest or shoulders	The method is unfavorable as the sensors are attached directly to the ankles
Control and monitoring of rural healthcare [Wang and Cai 2020]	Gives trade-off among end-to-end delay, energy consumption, system throughput that makes system suitable for healthcare systems	Authorization and authentication of this method is not clear
Classification of heart sound [Redlarski et al. 2014]	Used for smooth communication	The method doesn't give accuracy in case of distortion
Stress detection [Bharathi 2020]	Contributing the development of machines that helps human to understand human mental states	The real time ML techniques need to give quick respond so that the drivers are able to know the health status
Cardiac health monitoring [Zhang et al. 2021]	Delivers accurate information to doctors to improve disease diagnosis	Labeling process is very complex and consumes more time
Classification of breast cancer [Khan et al. 2019]	Gives more accurate data in the detection of breast cancer and reduces computational cost	DL algorithm needs to validate the precision

REFERENCES

Akmandor, A. O., and Jha, N. K. 2017. Smart health care: an edge-side computing perspective. *IEEE Consum. Electron. Mag.* 1: 29–37.

Al-Dweik, A., Muresan, R., Mayhew, M., and Lieberman, M. 2017. IoT-based multifunctional scalable real-time enhanced road side unit for intelligent transportation systems. In *IEEE 30th Canadian Conference on Electrical and Computer Engineering*, 1–6. IEEE, Windsor, ON, Canada.

Alahakoon, D., Nawaratne, R., and Xu, Y. 2020. *Self-Building Artificial Intelligence and Machine Learning to Empower Big Data Analytics in Smart Cities. Information Systems Frontiers*. https://doi.org/10.1007/s10796-020-10056-x.

Alahi, M. E. E., Nag, A., Mukhopadhyay, S. C., and Burkitt, L. 2018. A temperature-compensated graphene sensor for nitrate monitoring in real-time application. *Sensors and Actuators* 269: 79–90.

Arias, O., Wurm, J., Hoang, K., and Jin., Y. 2015. Privacy and security in internet of things and wearable devices. In *IEEE Transactions on Multi-Scale Computing Systems,* 1(2): 99–109, 1 April–June 2015. DOI: 10.1109/TMSCS.2015.2498605.

Bharathi, R., Abirami, T., Dhanasekaran, S., Gupta, D., Khanna, A., Elhoseny, M., and Shankar, K. 2020. Energy efficient clustering with disease diagnosis model for IoT based sustainable healthcare systems, Sustainable Computing. *Informatics and Systems* 28.

Cabaj, K., and Mazurczyk, W. 2016. Using software defined networking for ransomware mitigation: The case of CryptoWall. *IEEE Networks* 30: 14–20.

Castro-Neto, M., Jeong, Y. S., Jeong, M. K., and Han, L. D. 2009. Online-SVR for short-term traffic flow prediction under typical and atypical traffic conditions. *Expert Systems with Applications* 36: 6164–6173.

Chang, S. H., Chiang, R. D., Wu, S. J., and Chang, W. T. 2016. A contextaware, interactive M-health system for diabetics. *IT Professional*. 18(3): 14–22. doi: 10.1109/MITP.2016.48

Dey, S., Chakraborty, A., Naskar, S., and Misra, P. 2012. Smart city surveillance: Leveraging benefits of cloud data stores. In *IEEE Conference on Local Computer Networks Workshops*, 868–876. IEEE, Clearwater, FL, USA.

Greco L. et al. 2020. Trends in IoT based solutions for health care: Moving AI to the edge. *Pattern Recognition Letters* 135: 346–353.

Habibzadeh, M., Hassanalieragh, M., Soyata, T., and Sharma, G. 2017. Solar/wind hybrid energy harvesting for supercapacitor-based embedded systems. In *IEEE Midwest Symposium on Circuits and Systems*, 329–332. IEEE, Boston, MA, USA.

Habibzadeh, H., TolgaSoyata, C., Kantarci, B., and Boukerche, A. 2019. Smart City system design: A comprehensive study of the application and data planes. *ACM Computing Surveys*, 52(2): 1–38, https://doi.org/10.1145/3309545

He, W., Yan, G., and Xu, L.D. 2014. Developing vehicular data cloud services in the IoT environment. *IEEE Transactions on Industrial Informatics* 2: 1587–1595.

Healey, J., and Picard, R. W. 2005. Detecting stress during real-world driving tasks using physiological sensors. *IEEE Transactions on Intelligent Transportation Systems* 6: 156–166.

International Atomic Energy Agency. (1998–2019). "Agricultural water management," www.iaea.org/topics/agricultural-water-management.

Jeyaraj, P. R., and Samuel Nadar, E. R. 2019. Smart-Monitor: Patient Monitoring System for IoT-Based Healthcare System Using Deep Learning, *IETE Journal of Research*. 1–8.

Khan, S., Islam, N., Jan, Z., Din, I. U., Khan, A., and Faheem, Y. 2019. An e-Health care services framework for the detection and classification of breast cancer in breast cytology images as an IoMT application. *Future Generation Computer Systems* 98: 286–296.

Khan, S., and Thirunavukkarasu, K. 2019b. "Indexing issues in spatial big data management." In *International Conference on Advances in Engineering Science Management & Technology (ICAESMT)-2019*, 1–5. Uttaranchal University, Dehradun, India.

Khan, S., Qader, M. R., Thirunavukkarasu, K. and Abimannan, S. 2020. "Analysis of Business Intelligence Impact on Organizational Performance." In *2020 International Conference on Data Analytics for Business and Industry: Way Towards a Sustainable Economy (ICDABI)*, pp. 1–4. IEEE, Sakheer, Bahrain.

Li, H., Li, Y., and Porikli, F. 2016. DeepTrack: Learning discriminative feature representations online for robust visual tracking. *IEEE Transactions on Image Processing* 25: 1834–1848.

Liakos, K., Busato, P., Moshou, D., Pearson, S., and Bochtis, D. 2018. Machine learning in agriculture: A review. *Sensors* 18: 2674.

Luan, T. H., Gao, L., Li, Z., Xiang, Y., and Sun, L. 2015. Fog computing: Focusing on mobile users at the edge. *CoRR* abs/1502.01815 (2015).arxiv:1502.01815 arxiv.org/abs/1502.01815.

Manikandan, D., Manoj, A., and Sethukarasi, T. 2020. Agro-Gain - An Absolute Agriculture by Sensing and Data-Driven Through Iot Platform. *Procedia Computer Science* 172: 534–539.

Mehra, M., Saxena, S., Sankaranarayanan, S., Tom, R. J., and Veeramanikandan, M. 2018. IoT based hydroponics system using deep neural networks. *Computers and Electronics in Agriculture* 155: 473–486.

Miles, B., Bourennane, E. B., Boucherkha, S., and Chikhi, S. 2020. A study of LoRaWAN protocol performance for IoT applications in smart agriculture. *Computer Communications* 164: 148–157.

Mohammadi, M., Al-Fuqaha, A., Sorour, S., and Guizani, M. 2018. Deep learning for IoT big dataand streaming analytics: A survey. *IEEE Communications Surveys & Tutorials*. IEEE. DOI: 10.1109/COMST.2018.2844341

Muangprathub, J., Boonnam, N., Kajornkasirat, S., Lekbangpong, N., Wanichsombat, A., and Nillaor, P. 2019. IoT and agriculture data analysis for smart farm. *Computers and Electronics in Agriculture* 156: 467–474.

Occhiuzzi, C., and Marrocco, G. 2010. The RFID technology for neurosciences: Feasibility of limbs' monitoring in sleep diseases. *IEEE Transactions on Information Technology in Biomedicine*. 14: 37–43.

Page, A., Kocabas, O., Soyata, T., Aktas, M. K., and Couderc, J. 2014. Cloud-based privacy-preserving remote ECG monitoring and surveillance. *Annals of Noninvasive Electrocardiology*. 20: 328–337.

Pejovic, V., Musolesi, M. 2015. Anticipatory mobile computing: A survey of the state of the art and research challenges. *ACM Computing Surveys* 47: 1–29.

Rathore, M. M., Ahmad, A., Paul, A., and Jeon, G. 2015. Efficient Graph-Oriented Smart Transportation Using Internet of Things Generated Big Data. In *11th International Conference on Signal-Image Technology & Internet-Based Systems* (SITIS), 512–519. IEEE, Bangkok, Thailand.

Razzak, M.I., Naz, S., and Zaib, A. 2018 Deep learning for medical image processing: Overview, challenges and the future. In: Dey N., Ashour A., and Borra S. (eds.), *Classification in BioApps. Lecture Notes in Computational Vision and Biomechanics*, vol. 26, 323–350. Springer, Cham. https://doi.org/10.1007/978-3-319-65981-7_12.

Redlarski G., Gradolewski, D., and Palkowski, A. 2014. A system for heart sounds classication. *PloS One* 9: Art. no. e112673.

Satyanarayanan, M. 2017. The emergence of edge computing. *Computer* 50: 30–39.

Shwe, H. Y., Jet, T. K., and Chong, P. H. J. 2016. An IoT-oriented data storage framework in smart city applications. In *International Conference on ICT Convergence*, 106–108. IEEE, Jeju, Korea.

Siow, E., Tiropanis, T., and Hall, W. 2018. Analytics for the internet of things: A survey. *ACMComputing.Surveys* 51: 4.

Sladojevic, S., Arsenovic, M., Anderla, A., Culibrk, D., and Stefanovic, D. 2016. Deep neural networks-based recognition of plant diseases by leaf image classification. *Computational Intelligence and Neuroscience* 2016: 1–11.

Stojmenovic. I. 2014. Fog computing: A cloud to the ground support for smart things and machine-to-machine networks. In *Proceedings of the Australian Telecommunication Networks and Applications Conference* (ATNAC' 14). *IEEE*, 117–122.

Sundmaeker, H., Verdouw, C., Wolfert, S., and Perez Freire, L. 2016. Internet of food and farm 2020. *Digitising the Industry-Internet of Things Connecting Physical, Digital and Virtual Worlds*. Vermesan, O., and Friess, P. (Eds.), 129–151.

Tang, W., Zhang, K., Zhang, D., Ren, J., Zhang, Y., and Shen, X. S. 2019. Fog-enabled smart health: toward cooperative and secure healthcare service provision. *IEEE Communications Magazine* 57:42–48.

Wang, X., and Cai, S. 2020. Secure healthcare monitoring framework integrating NDN-based IoT with edge cloud, *Future Generation Computer Systems* 112: 320–329.

Wolgast, G., Ehrenborg, C., Israelsson, A., Helander, J., Johansson, E., and Manefjord, H. 2016.Wireless body area network for heart attack detection [education corner]. *IEEE Antennas and Propagation Magazine* 58: 84–92.

Wu, G., Chen, J., Bao, W., Zhu, X., Xiao, W., and Wang, J. 2017. Towards collaborative storage scheduling using alternating direction method of multipliers for mobile edge cloud. *Journal of Systems and Software* 134: 29–43.

Wu, Q., Ding, G., Xu, Y., Feng, S., Du, Z., Wang, J., and Long, K. 2014. Cognitive internet of things: A new paradigm beyond connection. *IEEE Internet Things* J:1129–143.

Yin, J., Gorton, I., Poorva, S. 2012. Toward real time data analysis for smart grids. In *Proceedings of the 2012 SC Companion: High Performance Computing, Networking Storage and Analysis.* IEEE, Salt Lake City, UT, USA, 827–832.

Zhu, N. 2015. Bridging e-health and the Internet of Things: The SPHERE project. *IEEE Intelligent Systems* 30: 39–46.

Zhang, D., Xia, Z., Yang, Y., Yang, P. O., Xie, C., Cui, M., and Liu, Q. 2021. A novel word similarity measure method for IoT-enabled Healthcare applications, *Future Generation Computer Systems* 114: 209–218.

Zhao, H., Chen, P.-L., Khan, S. and Khalafe, O. I. 2020. "Research on the optimization of the management process on internet of things (Iot) for electronic market." *The Electronic Library.*

5 Machine Learning and Deep Learning Are Crucial to the Existence of IoT and Big Data

Muhammad Tahir, Nawaf N. Hamadneh, and Mohammad Khalid Imam Rahmani

CONTENTS

5.1 INTRODUCTION

The Internet of Things (IoT) refers to billions of objects (things) that are connected to each other over the internet where they collect data from real-world observations and exchange that data with each other (Rayes and Salam 2017). This huge network is contributing to the production of a huge amount of data (big data) that needs more intelligent applications in the IoT paradigm (Fenila et al. 2021). Smart choices of ML and DL based algorithms can help decision makers in identifying important patterns

DOI: 10.1201/9781003138037-5

in big data (Al-Garadi et al. 2020; Khan and Kannapiran 2019). The selection of ML and DL techniques depends on the type and amount of data that can be processed promptly on time.

Machine learning (ML) is a sub-field of Artificial Intelligence (AI) that has been utilized to address a wide range of real-world problems (Awan et al. 2021; Tahir et al. 2020). The ML techniques have the capability of learning natural patterns in data using their features (Subasi 2020). ML enabled computers can perform tasks mimicking the abilities of humans. Usually, ML techniques can be used to develop models based on historical data, and then this model can be used to predict unseen future data (Subasi 2020). ML systems can be trained on existing/historical data after which they can improve and generalize their learning capabilities by acquiring new knowledge without developing new programs (Woolf 2010). The performance of ML algorithms is highly dependent on the amount of historical data on which they are trained. ML algorithms have a key role to play with the advent of big data where they are used for decision making in computational finance, energy production, image processing (Bashir et al 2017), computer vision, natural language processing (Shahnawaz and Mishra 2015; Khan et al. 2018), and stock trading. Supervised and unsupervised learning are the two popular approaches in ML where the former uses labeled data for training an ML algorithm whereas the latter adopts the notion of grouping data into different clusters based on similar patterns. Supervised learning is considered as classification when the output label is categorical. On the other hand, it is considered as regression when the output is continuous. Classification techniques include but are not limited to support vector machines, k-nearest neighbors, and neural networks. Regression can be performed using stepwise regression, bagging and boosting algorithms, and neural networks. Unsupervised learning refers to grouping data on the basis of hidden patterns without using the data labels. K-means, hierarchical clustering, hidden Markov models, and self-organizing maps are some of the examples of unsupervised learning. Deep Learning (DL) is a branch of ML that focuses on the applications of artificial neural networks such as in machine translation (Shahnawaz 2011), image processing, resource planning (Xiang et al. 2021), and so forth. In contrast to ML algorithms, DL based techniques learn from the raw data directly without pre-processing. DL techniques require a huge amount of data for training and are capable of learning complex patterns from the input data. Cloud service providers are offering ML enabled platforms where developers can apply ML techniques to big data for information extraction and analysis. This has become possible due to the integration of support for ML prediction systems.

ML and DL methods can be applied effectively to the data generated and collected by the things in the IoT and later on further utilized to identify hidden patterns and extract useful information to inform decision making. ML and DL techniques can guarantee privacy and security to the data on the cloud.

5.2 THE INTERNET OF THINGS (IOT)

'Things' in the realm of the IoT refers to physical sensors, actuators, or embedded systems that communicate over the internet to make informed and smart decisions collectively and, in a timely manner. The objective of IoT infrastructure is to enable the things to exchange information and make informed decisions autonomously, and with perfection, but without human intervention. The IoT is playing a key role in businesses, health monitoring services, and civilian security. Since IoT enabled devices are supposed to collect data, perform machine to machine communication, and execute simple pre-processing algorithms, it follows that IoT systems should be capable of completing these tasks with minimal computational power as well as reduced power consumption, at a low cost. The IoT provides connectivity to physical devices over the internet that enables the things to continuously generate, collect, and share data with each other. The big data generated in this way requires the IoT to be able to process and analyze this data. In order to perform different tasks remotely, IoT systems are equipped with monitoring and controlling capabilities (Syafrudin et al. 2018; Zhao et al 2020). Note that, processing and analysis of data are performed in the things and not in a centralized system. Monitoring and controlling capabilities have a huge impact on the wellbeing of consumers, effectiveness of governments, efficiency of healthcare service providers, and profitability of businesses.

5.3 IOT COMPONENTS

An IoT system consists of sensor, gateway, connectivity, cloud, analytics, and user interface (Rayes and Salam 2017). A sensor is responsible for collecting information from the external environment. Its job is to detect variations and to communicate those changes to the cloud. The gateway manages the flow of data and offers encryption services for the data to prevent unauthorized access to the data. Connectivity is a key factor for establishing communication among IoT devices. The IoT cloud offers storage services to the huge amount of data generated by IoT things. The cloud provides analytical tools for further analysis of the data that is accessible through the internet. The changes are captured by the IoT analytics, which processes the data and presents it to the user for further analysis. The user interface enables interaction with the system where the user can visualize the data and information, respond to the triggers, or act upon the various notifications.

5.4 IOT FEATURES

The characteristics of the IoT are what played a vital role in the widespread acceptance of the IoT and include connectivity, analyzing, integrating, artificial intelligence, sensing, active adherence/active engagement, and endpoint management.

Connectivity is the primary feature that connects the IoT things to the cloud over the internet. Reliable, secure, and bi-directional communication depends on the high-speed communication links between the things and the cloud. Analyzing refers to real-time analysis after data collection through connectivity, and is further used to make informed and intelligent decisions. Integrating, refers to the integration of different models to enhance the user experience. IoT systems are accepted in the market due to their intelligent behavior. They can detect minute changes in the environment that are reported in a timely fashion for appropriate decision making. Sensing capability is provided with the availability of IoT sensor devices without which the IoT would not be able to detect and monitor any changes or variations in the environment. The IoT has transformed passive networks into active networks. In this connection, IoT infrastructure, services, and other resources engage each other actively. IoT endpoints refers to the things that collect data about a service, or machine for monitoring purposes and share it with the cloud for further analysis. Endpoint management is an important requirement for an IoT setup. Security is another important feature of an IoT system that guarantees the security of personal data collected over its infrastructure.

5.5 IOT ARCHITECTURE

IoT Architecture (Yaqoob et al. 2017) is composed of the following four stages.

- Things and sensors.
- IoT data acquisition systems and gateways.
- Edge devices.
- Cloud.

First, things or sensors are IoT devices that communicate with each other over the internet and are responsible for sensing the data and information from the environment, collecting them, and communicating that information with the IoT gateways. Next, the IoT data acquisition system and gateway pre-process the collected data for further analysis. Then, edge devices further process the data and provide advanced analysis. Finally, the data is sent to the cloud where ML and DL techniques may be used for advanced analytics and processing. The data is also shared with other devices for smarter and well-informed decision-making. Some of the popular applications of the IoT include smart homes, smart cities, healthcare monitoring, wearables, smart transportation, supply chain management, security, and surveillance systems.

5.6 ML AND DL TECHNIQUES

ML algorithms can identify/recognize important patterns in unseen data once they have been trained using historical data (Hamadneh et al. 2021). ML algorithms have to go through certain stages to train and test an algorithm. These stages include: pre-processing, feature extraction, feature selection, and classification/regression. While training an ML algorithm, a human researcher must monitor the process interactively particularly in selecting the appropriate algorithm for a particular problem domain.

DL algorithms on the other hand do not require these cumbersome steps. They are capable of processing raw data in an end-to-end fashion. The system can be fed with images or raw data directly to where they will be processed, and the final output will be computed without human involvement. There is no need for explicit feature extraction and selection. All the tasks are automatically handled in the layers of the DL process where 'deep', in this case, indicates the number of extra hidden layers. This contrasts with 'shallow' networks. DL algorithms are computationally intensive, requiring powerful CPUs, and GPUs along with a huge memory capacity.

DL is a branch of ML that has demonstrated outstanding performance in the designing of autonomous cars for recognizing traffic signs and pedestrians, speech analytics, image classification, and many more. All this is achieved directly from raw images or unstructured data without human intervention.

5.7 THE ROLE OF ML AND DL TECHNIQUES IN IOT

ML based IoT solutions are increasingly gaining the attention of researchers, who, in the recent past have demonstrated their importance in our daily lives (Adi et al. 2020; Song et al. 2018). ML and DL enabled IoT has equipped our physical world with a digital nervous system. ML and DL techniques are required to analyze and extract useful patterns from the big data generated from billions of devices connected to the IoT.

ML and DL algorithms play a key role in turning these devices into smarter things over IoT. The efficiency of IoT can be enhanced using ML and DL techniques by analyzing the big data in a timely manner. Similarly, intrinsic patterns can also be identified by using decision trees, neural networks, and clustering techniques to name a few from ML and DL theory. It is evident that ML and DL algorithms act as a core requirement to enable IoT devices to make intelligent and informed decisions independently without human intervention. However, data collection and information extraction may be monitored by human observers in some cases for intelligent application development. However, IoT systems should be capable of making intelligent decisions independently using the collective knowledge of the things in the system. Intelligent behavior of IoT systems is also expected in the development of IoT applications where resources are allocated automatically and optimally.

ML techniques can not only be applied for the optimization of the IoT infrastructure but also for data analysis and decision making. For example, ML can optimize resource allocation, reduce congestion, and manage other network parameters to help establish a strong and efficient infrastructure. Similarly, ML can be used to pre-process the data, analyze it to identify intrinsic patterns and visualize them for a human decision-maker. Another important consideration is the utilization of DL techniques to handle big data since IoT infrastructure allows the addition of an avalanche of new devices. The term big data here refers to the reality that a huge amount of structured and unstructured data cannot be handled by traditional relational databases. Therefore, with enhanced IoT infrastructure, DL techniques can be effective in inferencing and extracting hidden information, making it easier for decision-makers to make smart and informed decisions.

5.8 IOT APPLICATIONS

IoT has found many applications in our society including, but not limited to, security and surveillance systems, healthcare monitoring systems, businesses, smart homes, smart transportation, and smart cities (Bhattacharya et al. 2020). Individuals can use IoT systems to monitor their daily life routines, manage their reservations in restaurants, and receive notifications related to different services. When it comes to businesses, IoT can efficiently monitor stock levels in warehouses, manage the supply chain, and perform predictive maintenance. Some popular and widespread applications are discussed below.

5.8.1 IoT in Agriculture

In order to meet the growing food requirements, the IoT (Elijah et al. 2018) can play a key role in increasing the production of crops. In this context, the nature of the soil has a great effect on the production of crops. IoT sensors can be used to detect the soil condition and make this information available to the farmers who can use this information for decision making. The sensors can detect moisture, acidity, temperature of the soil under different weather conditions. This will help farmers to plan irrigation and sowing related decisions effectively.

5.8.2 IoT in Healthcare

Patients admitted to hospital, particularly to intensive care units need constant monitoring by the healthcare providers. IoT enabled sensors can offer this service that can not only monitor the health condition of a patient in a hospital bed but also in a remote location and in real-time (Zeadally and Bello 2019). Likewise, hospital beds can be transformed into smart beds with IoT technology, which can communicate the blood pressure, body temperature, and other similar measures of a patient's body function to the doctors, in a timely manner.

5.8.3 IoT in Transportation

IoT systems are also demonstrating their feasibility in transportation (Kumar and Dash 2017) by allowing businesses to identify shorter routes in terms of reduced fuel costs or time costs for their cargo shipments. Similarly, the data collected from different sensors around a city can also be used to monitor traffic patterns and the usage of parking spaces during different hours of the day. This will help city planners to design traffic policies that meet the needs of the public.

5.8.4 IoT in Government

Government authorities can use IoT (Wirtz et al. 2019) sensors to collect data across multiple sectors that can be used to provide better governance. IoT based monitoring systems autonomously perform their tasks and they outperform conventional monitoring systems. Having all the information to hand, authorities can come up with well-informed and timely decisions, meeting the larger needs of the public.

5.8.5 IoT in Energy

Several countries are facing energy crises around the world. The IoT has great potential for the regulation of energy consumption and can result in energy saving. IoT sensors coupled with machine learning algorithms (Hossain et al. 2019) can collect data about line losses, energy consumption, intrusion, and electricity usage patterns in different weather conditions. Similarly, gathering temperature and other environmental conditions may also help in regulating electricity usage. The data collected by IoT sensors can be analyzed for informed decision-making to balance the production and consumption of energy.

5.8.6 IoT in Homes

The dream of having smart homes has come true with the advent of IoT, which has enabled homeowners to monitor IoT based home appliances from remote locations. IoT based smart homes (Gaikwad et al. 2015) are capable of performing different tasks, for instance, turning on/off an air conditioner in response to weather conditions. Similarly, the face of an entrant can be recognized when a door is opened. Some other utility examples include but are not limited to ventilation, management of water temperature, surveillance, and the switching on/off of lights on exit/entry to a room.

5.8.7 IoT in the Supply Chain

IoT has enabled many businesses (Manavalan and Jayakrishna 2019) to introduce novel business processes that are more efficient and flexible in nature. IoT embedded sensors enabled business owners to monitor their products from production to delivery. The data collected through IoT sensors during these processes provides a basis for well-informed decision-making that leads to enhanced customer experience and optimized production.

5.9 FUTURE RESEARCH POTENTIAL

IoT enabled technologies are gaining widespread global acceptance. The IoT has the potential to generate $11.1 trillion each year by 2025 (Manyika et al. 2015). The IoT is offering much more to individuals and commercial businesses in terms of control over day-to-day tasks. It has already been integrated with big data and it has a great potential for blockchain applications. It is evident that the IoT is changing our economic, commercial, and social values. The continuous communication of IoT sensors with each other and the cloud do, however make it vulnerable to security threats. Therefore, new cryptographic techniques are needed to address these problems since the existing techniques are not capable of addressing issues related to IoT infrastructure. ML/DL techniques can deliver the required solutions to the IoT systems that can use them, to know about the hidden and intrinsic patterns in big data.

5.10 CONCLUSION

The internet of things is gaining the attention of many academic researchers as well as commercial entities, who are working on it to capture the market. Due to the advancement in smart hand-held devices, it has now become feasible to process the data locally without involving the cloud. In this chapter, we have identified the significance of ML/DL techniques to be integrated with IoT and cloud for making them smarter and more effective in timely decision making. Since the data generated by IoT sensors is getting huge and transforming into big data, therefore, ML/DL techniques will have potential applications in transforming this data into well-understood information.

REFERENCES

Adi, Erwin et al. 2020. "Machine learning and data analytics for the IoT". In: *Neural Computing and Applications* 32, pp. 16205–16233.

Al-Garadi, Mohammed Ali et al. 2020. "A survey of machine and deep learning methods for internet of things (IoT) security". In: *IEEE Communications Surveys & Tutorials* 22.3, pp. 1646–1685.

Awan, Nabeela et al. 2021. "Machine learning-enabled power scheduling in IoT-based Smart Cities". In: *Computers, Materials & Continua* 67.2, pp. 2449–2462.

Bashir, Tariq, Imran Usman, Shahnawaz Khan, and Junaid Ur Rehman. 2017. "Intelligent reorganized discrete cosine transform for reduced reference image quality assessment." *Turkish Journal of Electrical Engineering & Computer Sciences* 25.4, pp. 2660–2673.

Bhattacharya, Sweta, Somayaji, S.R.K., Gadekallu, T.R., Alazab, M., and Maddikunta, P.K.R. 2020. "A review on deep learning for future smart cities". In: *Internet Technology Letters*, e187.

Gaikwad, Pranay P., Jyotsna P. Gabhane, and Snehal S. Golait. 2015. "A survey based on Smart Homes system using Internet-of-Things". In: *2015 International Conference on Computation of Power, Energy, Information and Communication* (ICCPEIC). IEEE, Melmaruvathur, India pp. 0330–0335.

Hamadneh, Nawaf N., Muhammad Tahir, and Waqar A. Khan. 2021. "Using artificial neural network with prey predator algorithm for prediction of the COVID-19: The case of Brazil and Mexico". In: *Mathematics* 9.2, p. 180.

Hossain, Eklas et al. 2019. "Application of big data and machine learning in smart grid, and associated security concerns: A review". In: *IEEE Access* 7, pp. 13960–13988.

Khan, Shahnawaz, Usama Mir, Salam S. Shreem, and Sultan Alamri. 2018. "Translation divergence patterns handling in English to Urdu machine translation." *International Journal on Artificial Intelligence Tools* 27.05: 1850017.

Khan, Shahnawaz, and Thirunavukkarasu Kannapiran. 2019. "Indexing issues in spatial big data management." In: *International Conference on Advances in Engineering Science Management & Technology (ICAESMT)-2019*, Uttaranchal University, Dehradun, India.

Kumar, Nallapaneni Manoj and Archana Dash. 2017. "Internet of things: an opportunity for transportation and logistics". In: *Proceedings of the International Conference on Inventive Computing and Informatics* (ICICI 2017), *Coimbatore, India*, 23rd to 24th November 2017, pp. 194–197.

Manavalan, E. and K. Jayakrishna. 2019. "A review of Internet of Things (IoT) embedded sustainable supply chain for industry 4.0 requirements". In: *Computers & Industrial Engineering* 127, pp. 925–953.

Manyika, James et al. 2015. Unlocking the potential of the Internet of Things. URL: www.mckinsey.com/business-functions/mckinsey-digital/our-insights/the-internet-of-things-the-value-of-digitizing-the-physical-world (visited on 2021).

Naomi, J. Fenila, M. Kavitha and V. Sathiyamoorthi. 2021. "Machine and deep learning techniques in IoT and cloud". In: *Challenges and Opportunities for the Convergence of IoT, Big Data, and Cloud Computing*. IGI Global, pp. 256–278.

Olakunle Elijah et al. 2018. "An overview of Internet of Things (IoT) and data analytics in agriculture: Benefits and challenges". In: *IEEE Internet of Things Journal* 5.5, pp. 3758–3773.

Rayes, Ammar and Samer Salam. 2017. "Internet of things from hype to reality". Springer.

Shahnawaz, M. R. 2011. "ANN and rule based model for English to Urdu-Hindi machine translation system." In *Proceedings of National Conference on Artificial Intelligence and agents: Theory& Application*, AIAIATA, Varanasi, India, pp. 115–121. 2011.

Shahnawaz, and R. B. Mishra. 2015. "An English to Urdu translation model based on CBR, ANN and translation rules." In: *International Journal of Advanced Intelligence Paradigms* 7.1 pp. 1–23.

Song, Mingcong et al. 2018. "In-situ ai: Towards autonomous and incremental deep learning for IoT systems". In: *2018 IEEE International Symposium on High Performance Computer Architecture* (HPCA). IEEE, Vienna, Austria, pp. 92–103.

Subasi, Abdulhamit. 2020. *Practical Machine Learning for Data Analysis Using Python*. Academic Press, USA.

Syafrudin, Muhammad et al. 2018. "Performance analysis of IoT-based sensor, big data processing, and machine learning model for real-time monitoring system in automotive manufacturing". In: *Sensors* 18.9, p. 2946.

Tahir, Muhammad et al. 2020. "Discrimination of Golgi proteins through efficient exploitation of hybrid feature spaces coupled with SMOTE and ensemble of support vector machine". In: *IEEE Access* 8, pp. 206028–206038.

Wirtz, Bernd W., Jan C. Weyerer, and Franziska T. Schichtel. 2019. "An integrative public IoT framework for smart government". In: *Government Information Quarterly* 36.2 (2019), pp. 333–345.

Woolf, Beverly Park. 2010. *Building Intelligent Interactive Tutors: Student-centered Strategies for Revolutionizing e-Learning*. Morgan Kaufmann, USA.

Xiang, Xiaojun, Qiong Li, S. Khan, and Osamah Ibrahim Khalaf. 2021. "Urban water resource management for sustainable environment planning using artificial intelligence techniques." In: *Environmental Impact Assessment Review* 86: 106515.

Yaqoob et al. 2017. "Internet of Things Architecture: Recent Advances, Taxonomy, Requirements, and Open Challenges". In: *IEEE Wireless Communications* 24.3, pp. 10–16.DOI:10.1109/MWC.2017.1600421.

Zeadally, Sherali and Oladayo Bello. 2019. "Harnessing the power of Internet of Things based connectivity to improve healthcare". In: *Internet of Things*, vol. 14, p. 100074.

Zhao, Haoyu, Pei-Lin Chen, S. Khan, and Osamah Ibrahim Khalafe. 2020. "Research on the optimization of the management process on internet of things (IoT) for electronic market." *The Electronic Library*.

6 Design of a Novel Task and Update-Based Social App

Proof of Concept for Richer UI/UX

Suja Panicker, Sachin Vahile, Adrija Guin, and Rahul Sethia

CONTENTS

DOI: 10.1201/9781003138037-6

6.1 INTRODUCTION

The user interface (UI) is a medium to transmit and trade data between people and devices, and furthermore a coordinated operation framework for users to work on their devices. The UI serves as the entire framework for the end users of the software framework, thus satisfaction or disappointment in the prototype is primarily identified with nature of the UI design (Nurgalieva, Laconich et al. 2019). Attractiveness is not the only purpose of the UI but intuitiveness of the design is an important aspect as well (Zhafirah, Hardianto et al. 2019). The UI build should only be concerned with the target audience. More instinctive interfaces can at that point better support the utilization of such gadgets by more adults, if the target audience includes adults, subsequently incrementing their admittance to advanced items and e-administrations (Zhafirah, Hardianto et al. 2019). User experience tries to encourage rich, appealing interplay between users and frameworks. For this experience to unfurl in a cohesive sense, the user must feel inspired to increase their involvement with the gadget (Joo 2017).

6.2 MOTIVATION

In general, human activities can be as varied as the number of spices available in the market or as confusing as the number of articles of clothing that one has to go through before selecting and finalizing a particular article. Even if the selection has been finalized, there always exist some more shops specializing in the clothing choices that the person might opt for. At the same time, the increase in population has increased the number of questions asked daily. To answer every question is not possible for everyone. To avoid this problem from growing exponentially in the future, an idea has been developed that is based on a task and update-based project. The platform created in this project deals only with people's day to day queries.

Despite active research on this issue, the following problems remain unanswered.

- Firstly, platforms where you can interact as well as get a comprehensive answer to your questions from an expert in a particular field are scanty.
- Secondly, information available on different websites cannot be trusted blindly all the time.

In this paper a task and update-based platform is proposed where each question indicates an individual's task that needs to be completed and update signifies the replies made by other individuals present on the platform. Anyone with adequate knowledge regarding the task can update. This task can vary from basic needs to getting a large project underway. This platform can provide the required answers to all the user's queries and questions. With this work, we expect a seamless experience for the user who can ask anything without being judged and obtain the required information from an expert very quickly along with minimum irrelevant information.

Improvement in social perception by the general population can also be anticipated, thanks to familiarity with smart applications by this system. By being user-friendly and navigation-friendly, the design of itself simplifies the whole process of posting a task, getting corresponding updates, and careful design of other features such as: saved to archives, forwarding searching, and so forth. In this paper not only were people surveyed about the difficulties of collecting information for any kind of project but also the aim was to try to incorporate user experience (UX) principles as to how the whole experience of questioning and answering could be simplified as much as possible along with highlighting the updates received. The credibility of each individual's reply is verifiable from their profile.

The remaining section of the work is organized as follows: Section 6.3 is a literature review, Section 6.4 presents the technological process and platform used, Section 6.5 covers proposed work, Section 6 presents the experimental results toward proof of concept, Section 6.7 details the design and analysis, Section 6.8 presents novelty, Section 6.9 covers the research contribution, Section 6.10 highlights future work, and finally there is the conclusion.

6.3 LITERATURE REVIEW

In this section important domain knowledge pertaining to UI/UX is presented, along with common misconceptions between UI and UX, fundamental color models and color theory and summarized highlights of current research in the field (and gaps in current research).

An excellent, user-friendly interface can give an individual a pleasant experience. It will narrow the gap between the person and the electronic device. Interface design is pure art design. The UI design and the research included is intended to enhance the end users' visual experience (Fu 2010).

There are many sections of a UI design. A good UI design needs talent followed up with years of practice. As people try to go into the field of UI, it can become overwhelming because of the large amount of content that is available in papers, articles, and on the internet. A good UI design is not only eye catching but tends to engage the user with current trends and engages their emotions whilst they are interacting with it. The development process is based on market survey and research done using the general user as its focus for data collection. This has the potential for deciding who are likely to be the users of the application, the brand image that is to be portrayed and allows the creating of a roadmap to the process of designing.

These prototypes need to be tested. Tests are carried out to measure their feasibility. A usability test is a good way to test the functioning of the product or service. Usability estimation is completed by utilizing a sequence of questionnaires that can elicit information related to adequacy, proficiency, and fulfillment in the utilization of a data framework. One of the questionnaires set that can be utilized to gauge convenience is USE (Usefulness, Satisfaction and Ease of use), since it can cover the three components of usability estimation as indicated by ISO (Goel and Goel 2016).

The USE questionnaire is a poll that can be used for a usability questionnaire that is firmly linked to user experience. The results demonstrate the presence of any connections between the parameters: ease of use, and usefulness, that influence the degree of satisfaction (Goel and Goel 2016).

6.3.1 Framework of UI/UX Process

The popular framework, 5S is used. The product to be developed is segmented in the form of layers. This includes understanding the user's purpose and defining the perceptive and receptive attributes that will affect user behavior. One such framework of UX, the 5S, divides it into Strategy, Scope, Structure, Skeleton, and Surface. The interface is prioritized in such a manner that it emphasizes user needs first, which is the user's action plan, followed by the goals as a business project. The next step is to reflect on the results in order to find solutions to the problems that arise from the problem statement. This results in the formation of a prototype. Figure 6.1 shows the UI/UX interface thought diagram.

The process of designing begins with six stages of designing as discussed below (YouTube. Steps in the design process. 2018), Hungreebee Technologies 2021).

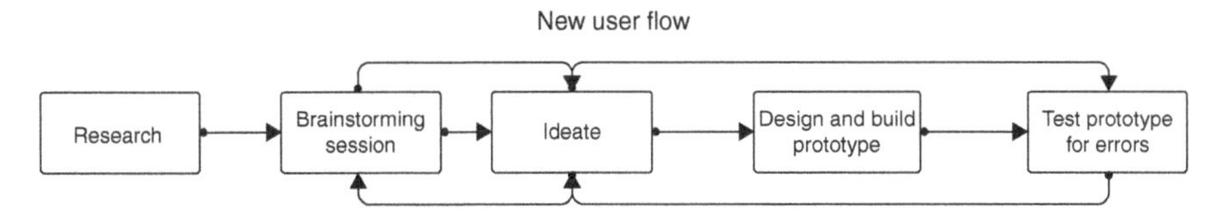

FIGURE 6.1 UI/UX interface thinking (Brien 2010).

6.3.1.1 Research

Doing a survey based on a pre-prepared questionnaire that is designed to collect as much relevant data as possible is important as a first step. The survey can function with a limited sample size that is carefully targeted to understand the problem statement and the brand image that is to be created. The candidates can be interviewed face to face or can be located online. The purpose of the survey and interview is to understand the daily life of the user and where the app or the product to be created will fit in. Along with this, new apps from Playstore can be reviewed and screenshots collected that deal with similar projects.

6.3.1.2 Sketches and IA

Information Architecture (IA) describes the layout of information and the flow of the application. The best way to start with IA is to make a flowchart depicting the product flow. After deciding the flow, the next step is to get the idea on paper in the form of rough sketches. Starting with design tools immediately is a big mistake, which can be overwhelming at this stage due to the number of options available. Thus, the foundation needs to be built on components. Components such as buttons and text areas are drawn on paper to decide their position on the screen. Drawing on paper provides a good visual map, trying out multiple options more quickly than on a platform.

6.3.1.3 Block Level Design

A decision is to be taken on the blocks of text, buttons, images, and other elements that are to be incorporated into the design. Each component is assigned a particular region. The user flow is decided for each component and its actions.

6.3.1.4 Wireframes

The next process is to draw the screens on paper and add in the minute details. These details might be a logo, icons, illustrations, text style, the map, comments about animation that are to be added, definitions of text, titles and subtitles. Screenshots taken during research should be used as references. The process can be started with pen and paper or on in a software package. Once the wireframes are drawn, feedback can be taken from potential users to get better structure, layout, and flow. Software packages can also be used but can restrict the flow of the project. Packages such as Balsamiq have been used for a decade. Other tools like Sketch and Figma may also be used.

6.3.1.5 Visual Design

Visual design (VD) refers to deciding what the product will finally look like to users. In this stage the look of each component is decided. This is the longest part in the designing stage. The available options at this stage can be reduced by focusing on factors which decide the brand image. This can be decided while wireframing.

6.3.1.6 Prototyping

The wireframes which serve as a blueprint for the product are converted into a mockup UI. The UI is converted into a prototype by linking screens together. The clickable prototype feels real and can used to get better feedback than using wireframes. This gives a more realistic effect. Adobe XD, Figma, Sketch, InVision, Gravit and Framerare are tools that are commonly used since they facilitate an easy design process and they provide multiple options to choose from. Prototyping tools exist that are made solely for this purpose, for example, ProtoPie and Flinto. Finally, the product is "handed off" to the developers. It is a collaborative process between the developers and UI/UX designers.

6.3.2 UI and UX are Discrete Stages

6.3.2.1 UI

A user interface (UI) alludes to a framework and a user connecting with one another through a protocol that operates the framework, inputs information and utilizes the contents (Brien 2010). This might incorporate the presentation screen, console, mouse, and desktop. This is how users communicate with applications or sites. This increased reliance leads numerous organizations that heavily utilize web applications and portable applications to give a greater importance to the UI so as to improve their organization's UE (User Equipment) (Nurgalieva, Laconich et al. 2019).

6.3.2.2 UX

User experience (UX) alludes to the general experience identified through the observations, feelings, thoughts, and responses of the user when directly or indirectly involved in the utilization of a framework, item, substance or administration (Brien 2010). It is the artistry of product design that encourages an interaction with finished products that will be as extraordinary as can be expected under the circumstances. This incorporates the association with the end-user on a few IT frameworks including the interface, illustrations, and plan. The two components are significant for product and task together (Nurgalieva, Laconich et al. 2019).

6.3.3 Misconceptions about UI/UX

There are common misconceptions about UI/UX (4 stages of UI Design. YouTube. 2018)

6.3.3.1: UX and UI are certainly different, however they are both linked, and both play their part. UX is to a large degree is a superset of UI. It is essential when designing an interface.

6.3.3.2: User experience isn't restricted to computerized innovations but spreads into various aspects of a product, including the product's life cycle. It is an iterative cycle that works around the whole product experience.

6.3.3.3: Design is not a single term, but an umbrella term that incorporates elements from different fields, for example, brain science, data analytics, innovation, user journey planning, and more.

6.3.3.4: Design isn't optional when developing a product from scratch. It is crucial to take care of u ser and business issues and fill in the void between business objectives and consumer contentment.

6.3.4 The Basic Principles of Designing

Barua outlines the basic principles of designing (2019).

6.3.4.1 White Space

Also referred to as negative space, it is the empty space between the elements in the user interface. Maintaining an appropriate proportion of white space among the components of the UI is very important not only for aesthetic purposes but also for ease of viewing the content.

6.3.4.2 Alignment

Alignment in the UI is the process of ensuring that every element is positioned correctly in relation to the other elements. Each element in a UI defines a series of rows and columns. Alignment of components used along a specific and unique row or column gives a much better effect.

6.3.4.3 Contrast

Contrast might be defined as being strikingly different from something, for example, using two or more colors that differ from each other. Naturally, every element in a UI has a certain amount of contrast based on the background. According to WCAG 2.0 Contrast Guidelines, Enhanced contrast, version (AAA) means that the visual presentation of text and images of text that has a contrast ratio of at least 7:1, except for large text which should have a contrast of at least 4.5:1. There are contrast checking tools available like browser plug-ins for Chrome and Firefox, websites where different values may be entered and UI design application plug-ins such as Sketch and Figma.

6.3.4.4 Scale

Along with the other fundamentals mentioned so far, the size of every UI element must be carefully considered. The scale of each element on the screen should be suitable to give a symmetrical yet aligned view. The white space shouldn't be overwhelming due to selection of the wrong scale.

6.3.4.5 Typography

Good typography requires an understanding of other fundamentals, along with a few special considerations. These considerations are:

6.3.4.5.1 Font Choice(s): It is advisable to avoid more than two font families. Most products use only one font family with which they commenced their project.

6.3.4.5.2 Visual Hierarchy: This is important as it is a way to establish the order of importance through typography. To establish this within a particular element, other fundamentals like font size, color, and contrast, and so forth. are manipulated during the setting of the typography.

6.3.4.5.3 Alignment: Alignment of all the text along a particular row or column or along a particular side (left, center or right when measured vertically or top, middle, or bottom if measured horizontally) can help in understanding the hierarchy.

6.3.4.5.4 Letter Spaci ng and Line Height: Proper spacing between components can increase readability. Spacing between letters also needs consideration according to the style selected.

6.3.4.5.5 Font Styles: The styles used need to be in harmony with the main theme of the project. Harmony of the product representation and the text presentation is important to portray a certain idea.

6.3.4.5.6 Color and Contrast: The difference in color between a title and a subtitle can provide the user with a good level of readability and can assist with easy navigation. Contrast helps in differentiating the different purposes of different parts of the text.

6.3.4.6 Color

The first UI design fundamental that shapes a user's experience is color. Decisions need to be made based on the psychology of color perception. Each color has a certain meaning to a certain group of people, for example, green is associated with wealth, growth, and nature whereas black can be used to signify luxury, sophistication, and elegance. One noteworthy observation is that depending on the company or the target audience, certain colors can convey different meanings to different communities and cultures. What the business wants to project should be considered, in the demo, in terms of eliciting emotions in the target audience. Before beginning to design, it is recommended that the color theme is decided upon and maybe that an advantage could be to reduce the colors that will be available in the palette. This might help to preclude the use of excessive ranges of color. Usually, the color palette should contain two colors, or a maximum of three colors. Too many different colors can destroy the quality. Often, a single color is used with different hues, tints or shades to sensitively reinforce different sections in the layout.

6.3.4.7 Visual Hierarchy

Every element of a user interface has a particular level of importance. Some elements are more important than others. Visual hierarchy is the way in which we establish this importance. For example, on a screen, if we lay out four boxes of the same size and at the same level, it will indicate to the user that all the blocks have equal importance. The moment one of the blocks is given a higher position on the screen, the eye automatically gravitates to that block first and then the remaining blocks are noticed. Now, if all of the blocks are given a darker shade to contrast with one white block on the same level of same size, that block will catch the user's eye first even if all of the blocks are on same level.

Other features include: closure, proximity, and similarity. Closure states that our brain tries to recognize any complex structure based on previous experience, even if the structure has some missing parts. Proximity which states that things that are close together appear to be more related than things that are at a distance from each other. Similarity states that objects sharing visual properties (size, color, texture, dimension, shape, or orientation) appear to be related and this is helpful in creating hierarchy (Vaniukov 2019).The UI fundamentals such as contrast, scale, or a combination of these, help to create a visual hierarchy.

6.3.5 Color Theory

Captivating visual content can be created with the aid of color theory. There are fundamental rules and guidelines that surround color and its use for designing aesthetically pleasing visuals. An understanding of the basics enables the deconstruction of the logical structure of color, to create and use color palettes more advantageously. The result leads to evoking a particular emotion, vibe, or aesthetic. It provides a foundation to create solid visuals (Soeegard 2020) as explained below.

6.3.5.1 Primary Colors

These are comprised of red, yellow, and blue separately. These hues are basic and cannot be created by mixing any other color. They are present in the overall design plan.

6.3.5.2 Secondary Colors

These colors are a mix of any two primary colors. There are three secondary colors: orange, purple, and green. These colors are in their purest form just as the colors from which they are derived from are in their purest form. Tints and shades can provide further variety.

6.3.5.3 Tertiary Colors

Tertiary colors are derived by combining a primary color with a secondary color that comes next to it on the color wheel.

6.3.6 Color Theory Wheel

The rationale behind the color wheel is discussed by Cartwright (2020) who is choosing color combinations by depicting how each color relates to the color that comes beside to it on a rainbow color scale. It is a circular graph that shows each primary, secondary and tertiary color, as well as their respective hues, tints, tones, and shades.

6.3.6.1 Hue

Hues are important considerations when combining two primary colors to create a secondary color. This is because a hue has the fewest options inside it. By mixing two primary colors that carry other tints, tones and shades, the result obtained will contain more than two colors.

6.3.6.2 Shade

A shade is the color obtained by adding black to any given hue. There are a wide range of possibilities.

6.3.6.3 Tint

Adding white to a color or hue, creates a tint. It is the opposite of shade.

6.3.6.4 Tone

Adding both white and black to a color creates a tone.

6.3.7 Additive and Subtractive Color Theory

Cartwright (2020) discussed two models that are generally used for color, CMYK, and the more popular RGB. These can be spotted in a computer program for designing, or in graphics.

6.3.7.1 CMYK

CMYK stands for cyan, magenta, yellow, and key. CMYK is the subtractive color model. This theory subtracts colors to get white. Basically, when you add up the colors cyan, magenta and yellow you get a black color. CMYK works on a scale of 0 to 100. If C = 100, M = 100, Y = 100 and K = 100 it gives a black color. If all values are equal to 0 then you end up with white. By adding colors, the white wavelengths are blocked from getting through. This model is generally used for ink cartridges in printers.

6.3.7.2 RGB

RGB stands for red, green, and blue and is based on the additive color model of light waves. The RGB color models are designed and intended for screen displays on electronic devices. Designing tools also use this model. This means, the more colors you add, the closer you get to white. It is created on a scale ranging from 0 to 255. Hence black is R = 0, G = 0 and B = 0 and white is vice versa. Many web programs will only give the RGB values or a HEX code.

6.3.8 Color Schemes

Cartwright (2020) explains that there are five color schemes followed. The selection of color that defines the context is to be considered before deciding the scheme. Color context refers to how we perceive colors as they contrast with one another.

6.3.8.1 Analogous Colors

Analogous color schemes are formed by pairing one main color with the two colors directly next to it on the color wheel. These structures do not create themes with much difference in contrast and are used to create a softer, less contrasting design.

For example, analogous structure can be used to create a color scheme with autumn or spring colors. This scheme can be used to create warmer or cooler color palettes that can serve the purpose of blending the elements in an image together.

6.3.8.2 Complementary Colors

Complementary colors provide the greatest amount of color contrast. This is possible as the colors selected sit directly across each other on a color wheel. The principal format is to base the design on one color and use the other as accents in the design. This color scheme is also great for displaying charts and graphs.

6.3.8.3 Monochromatic Color

Using this color scheme, one will be able to focus on one particular hue. It is based on various shades and tints of one hue. It lacks contrast but gives a polished and clean look. These colors do not pop out.

6.3.8.4 Triadic Colors

Triadic color schemes are derived by selecting three colors that are equally placed in lines around the color wheel. The three colors are equidistant from each other on the color wheel. This offers high contrasting colors while the tone remains the same. It can seem overwhelming if colors selected are on the same point on the line in a color wheel. This can be reduced by keeping a main color as the base color and using other two remaining colors sparingly or subdued by using a tint.

6.3.8.5 Split Complementary

This scheme includes one superior color, and the remaining two colors are directly adjacent to the superior color's complement. This is a kind of nuanced palette. This color scheme is difficult to handle as all the colors are meant to provide contrast. Though it provides great contrast, it might be tricky to strike the right balance.

Our observations from this study (Google 2020) are as follows:

- Some software automatically provides preset colors. Explore colors beyond the provided preset and decide on how best to use the colors for the desired design.
- Start with just one color and build the color scheme from there. To start a design with more than one color can be overwhelming and then later, it can become difficult to find the right harmony between the colors.
- Save the color schemes for later use. The color palette might suit a future design.
- Practice and play around more with colors. This brings in more expertise and builds skill.

6.3.9 Changes in UI/UX Trend Design Elements

Interfaces are firmly identified with design and connection (Brien 2010). Interface design assumes a significant part in connecting framework capacities. UX interface is additionally influenced by the ease of use of the framework, the user's knowledge and skills with the system.

6.3.9.1 Evolution of Minimal Design

The concept of minimalistic layout and minimizing complexity began appearing in the year 2017. The approach is to make the interface look as neat and clean as possible while minimizing the numbers components.

6.3.9.2 Moving Pictures

The most powerful sensation of the human body is vision. An image can speak more than a boring text would. A static image can describe an idea that would otherwise need many words. A dynamic image can explain an idea effectively where many hundreds of words would otherwise be needed. For an aging population, cognitive reaction-based intelligent UI/UX system can be optimized to give the best results for the senior population (Kumar, Kandaswamy et al. 2016).

6.3.9.3 Long Scrolling and Parallax Technique Website

The standardized format of most UIs is to create infinite scrolling. Some examples can be seen in newsfeeds or in websites meant for watching videos such as YouTube or social media apps like Instagram or Facebook.

6.3.10 Trends that Excite Modern Users

UX designs are saturated with fast-changing trends and tendencies (UI Design Crash 2019). However, there always exist long-term trends that define the industry. Modern companies create unique online experiences, fostering emotional connections with the community. Besides functionality, customers want new exciting experiences while interacting with brands in the digital world. It has led to the emergence of mobile apps and websites with the encompassing UI contents that unite surprising layouts, digital illustrations, and motion graphics, voice optimized search and much more.

6.3.10.1 Interactive UI

Mobile app and website animation in UI design has become very popular among users. Engaging animated stories, interacting with animated screens and funny animated heroes assisting users with helpful visual tips and explaining how to use products and services are some of the things users love (Kopf 2020). These visuals help to create trust with end-users, increase engagement, and provide great benefit for a customer support team that's overloaded with consumer questions. Elements like dynamic website backgrounds, kinetic typography, animated logos, and visual assistance, turn digital experiences into exciting adventures.

6.3.10.1.1 Illustrations

Illustrations are another popular digital trend used by companies across different industries such as Buffer, Pipedrive, and Boli. Google constantly partners with digital artists to produce creative illustrations for its products and services. Digital illustrations are the way to tell the brand story, brand mood and brand personality without using any words. Digital illustration in UI designs for mobile apps/websites helps establish emotional bonds with users and increases their loyalty.

6.3.10.1.2 Microinteractions

These are the little moments that stick to our memories and create an overall impression of things, events, and digital products. Microinteractions provide the opportunity to put a seed of excitement into a product UX and grow it into an unforgettable digital experience. Being almost invisible, they add dynamics, interactivity, and intuitiveness to UI design. The user should not get bored while waiting for a web page to be loaded. Every moment of user experience can be made sweet and delightful by fun-uploading, pull to refresh micro interaction, tab-animation, navigation micro interaction and more (UI Design Crash 2019).

6.3.10.2 Dark Mode

The dark UI is one of the hottest web design trends in 2020. Dark mode is considered to be one of the best UX practices as it minimizes user eye fatigue and facilitates scrolling an application or website with heightened comfort for eyes (Kim 2017).

6.3.10.3 Neumorphism

It seems more real than real. Neumorphism is a new take on skeuomorphism. Buttons, layouts, lights, cards, and other UI components look like real life objects. Neumorphism mimics reality and brings clean interfaces to life by adding physical elements and material design to a flat UI paradigm. Being just the reflection of real life, neumorphism turns mobile applications and web solutions into digital experiences that look like a part of our life (Rufo 2019).

We present highlights of current research in Table 6.1

6.4 TECHNOLOGICAL PROCESS AND PLATFORMS USED

6.4.1 Technological Process

6.4.1.1 Affirm Intended User

In designing the UI, the users who will be using the software will be defined by the characteristics of the software and its demand among them. The interaction used to develop the product should put emphasis on different possible users of the product. As an example, a particular font style might be used for old people while for young people a more modern look might be used.

6.4.2 Maintain Consistency

Maintaining consistency is discussed by Fu (2010).

6.4.2.1 Uniformity of Design Elements

The look of the software needs to be consistent and uniform across the whole product. There is no standard against which to measure it, hence the best way is to take user feedback.

TABLE 6.1
Highlights of Literature Survey

Reference and year published	Concepts learnt	Advantages
(Joo. 2017)	Changes in design trends, Modern trends, evaluation of understanding and analysis of UI/UX and mobile UI/UX build guide evaluation on subjects	Basic principles of designing, importance of wireframes and layout, process of research
(Zhafirah et al. 2020)	TCSD, usability, design of questionnaire for survey and evaluation of user requirements using likert system based on design principles, research methodology, user centered-requirement analysis using story board, conceptual model and wireframes and satisfaction and ease of use criteria for analysis	Focus on importance on wireframes, sketches and block level design, analysis based on other principles of design, prototyping and evaluation on prototype
(Zhao, Gao et al. 2012)	Consistency, usability and efficiency of design, UI research methodology, collection of reusable UI models, construction of framework at presentation layer and UI description language based on XML	Focus on service layer and effects used to engage the user and design trends followed
(Fu. 2010)	Principle of mobile UI design, process to confirm target user and guide user via the product, usability and Structure, Interactive and Visual design	Guidelines to follow to target a particular audience and increase engagement
(Kristiadi, Udjaja et al. 2017)	Difference between UI, UX and GX, survey to find target audience, use of statistics for analysis.	UI/UX of Tekken game interface and analysis of user audience and impact on audience.
(Almughram and Alyahya. 2017)	Agile process and User Centered Design (UCD), minimal design and frequent feedback, UX-designers and developers coordination, UCD-related activities, integration of UCD activities in distributed agile environments and agile project management tools	Limitations of the proposed system to enhance the integration of agile process into UCD.
(Weichbroth. 2020)	Evaluation of usability, analysis of usability attributes and methods to collect data	Proposed methods to increase the measure of lesser-known attributes like memorability, learnability and errors, more focus on simplicity and ease of use

(*continued*)

TABLE 6.1 (Continued)
Highlights of Literature Survey

Reference and year published	Concepts learnt	Advantages
(Fathauer and Rao. 2019)	Accessibility standards, agile development process	WCAG guidelines and comprehensive feedback on layout
(Brien. 2010)	User engagement with online shopping, Hedonic and Utilitarian Shopping Motivation Scale, User Engagement Scale and attributes of user engagement	Holistic approach to shopping online and balanced engagement
(Nurgalieva, Laconich et al. 2019)	Design guidelines for aging population	Feedback from aged audience based on the implementation, likert scale and design models to be followed
(Goel and Goel. 2016)	Concept of cloud, its use in e-commerce, cloud service models, deployment models, reduction cost on the hardware, backend and proposed model for an e-commerce product	Scalability of cloud, cloud model popular for e-commerce and other sources of direct and indirect income, integration of cloud into applications and affordability of cloud services among various groups of enterprises

6.4.2.2 Uniformity in Interactivity

The interactivity experienced by individual users must be consistent across various platforms. Different events trigger different emotions. User expected interactions should remain like the design.

6.4.2.3 Reduce Learning

Consistency limits the number of ways actions and operations are represented, ensuring that users don't have to learn new representations for each task. Further, establishing design norms like following platform conventions allow users to complete new tasks without having to learn a whole new toolset.

6.4.3 Platform Used

The platform used is an issue for consideration (Marketing Color Psychology. YouTube 2019), (Google. Neomorphism 2020).

Shifting to cloud-based computing (Cloud Computing benefits. Google 2020), (Verma 2019), (Bernal, Cambronero et al. 2019) has been the norm for a while. This migratory pattern has also been observed for design software which has resulted in the popularity of cloud-based apps like Figma. Along with multiple backups, cloud-based services offer encrypted security for data. This is something that is not always possible with local servers. Moreover, these services are always at a par with the latest innovations in the industry. This frees up the designer's time that would otherwise be spent on the up keep of the local systems. However, the main USP of cloud-based design apps is the wide-ranging opportunity for real time collaboration for the team. Now, the designers can focus on improving their designs whilst receiving feedback on the go. Recently, Zomato made the decision to switch to Figma to meet their design requirements. Given the new workplace standards that have emerged in recent times, cloud-based apps like Figma allow users to access their projects from any device, anywhere in the world. These developments, supported by the use of cloud-based software, result in richer innovation, increased efficiency, and better accessibility, all of which collectively improve the field of design.

Figma is an interface design application that runs in the browser. It's perhaps the best application for team-based collaborative design projects. Figma provides all the tools needed for the design phase of the project, including vector tools that are capable of fully-fledged illustration, as well as prototyping capabilities and code generation for hand-over.

We present highlights of various popular graphical cloud modeling languages in Table 6.2 (Varshney and Singh 2018).

In the proposed work, we have used Figma. However, as an extension of current work we propose to incorporate some graphical cloud-based modeling languages (shown in Table 6.2) that are based on the unified modeling language (UML).

Also, for the task-update based scheduling in current work, we propose to incorporate a combination of quality of service (QoS) features to facilitate optimization

TABLE 6.2
Highlights of Popular Graphical Cloud Modeling languages

Name	Brief description	Target
CAML	1) It characterizes the cloud-based topologies in UML suitable for deployment. 2) It considers virtual deployment target and components of the application.	IaaS, PaaS, SaaS
MULTICAP	1) It is based on a cloud-provider independent prospect. 2) Basic design is performed in UML. 3) Besides designing the multicloud applications as software artifacts, additional stereotypes (that allow annotation with QoS parameters, for example – response time) are provided. 4) It avoids cloud vendor lock-in.	IaaS, PaaS, SaaS
TOSCA	1) It characterizes portable automated deployment. 2) It emphasizes automatic management of the applications deployed on cloud.	IaaS, PaaS, SaaS
MOCCA	It migrates the existing software to cloud environment.	IaaS, PaaS, SaaS

in scheduling. Some features worth considering are: execution cost, deadlines, and execution time. We propose implementation of the novel vector evaluated particle swarm optimization (PSO) technique (Zhang, Liao et al. 2017) having the presence of two swarms (each one addressed to an objective), leading to a better collective solution set.

6.5 PROPOSED WORK

6.5.1 Research Goals

- A hassle-free experience while navigating through the product and not having to learn to use any new tools.
- A platform that is self-explanatory to post a task quickly.
- The option to save the tasks for later use.
- The option to delete the tasks whenever required.

6.5.2 Proposed Project

Important usage examples: login/sign up, post a task, update a task, search a task, and add a friend. Social media platforms are used for connecting, sharing information, building a circle of people with similar likes and dislikes, expressing emotions to promote oneself in our respective field of work. The UI is designed to keep people's minds engaged.

Problems encountered:

- These platforms are not primarily designed for general users to ask questions or to answer user questions.
- The feed might consist of videos and images irrelevant to user's tasks.
- Not many applications have been built specifically to answer users' queries based on day-to-day life.

6.5.3 Proposed Solution and System

6.5.3.1 Proposed Solution

We propose to build a UI of a social app for the company Hungreebee Technologies LLP (Hungreebee Technologies 2021) which will help users to get their queries answered by other knowledgeable people easily and quickly. It will help to keep all the answers collected in one place as saved updates. Hungreebee Technologies LLP is a start-up developed for the purpose of serving quality catering services for events, and which is now expanding to other fields such as the delivery of home cooked food as well as restaurant style food. It is one of India's largest catering aggregator platforms. Their aim is to provide an enriched catering experience to their users for events.

6.5.3.2 Proposed System

The designed system will give us a social media platform that overcomes the problems mentioned above. This system is designed on a task and update approach. A user posts a query, called a task and anyone from the circle can answer it, called an update (depending on the privacy turned on in the account by the user).

For example, a user wanted to know about how to get a birth certificate in Pune and posted a task for the same. The user specified their location to Pune. A user who is a part of the platform would update it. It is more likely that a user from Pune would be most likely to answer the question. The updates will be provided on the trending page where both activities, posting and updating, are possible. Distinction between tasks and updates will help the user view the updates at a glance. Social media is mostly used for leisure but this is not the case here. It is a knowledge and information-based platform and it has modes for maintaining privacy discussed below.

There are two modes in which the task can be updated.

6.5.3.2.1 Private Mode

In this mode the posted task will only be shown to the user's circle and only they can update the task.

6.5.3.2.2 Public Mode

In this mode the task will be visible to everyone though there will be an option to limit the updates from audience.

Once the users get the information they posted the task for, they can switch to a setting that will allow no more updates on that particular task however the updates that are already there can be read by other followers, and depending on the mode selected, the task can be hidden from the user's profile, deleted, or a limit can be placed on comments.

In the clickable prototype, a navigation bar is provided from where features can be accessed such as: home, notifications, searches for tasks, and a trending tasks page.

6.5.4 Research Methodology

Research methodology is discussed by Fu (2010), and Goel and Goel (2016).

Numerous researchers have talked about the significance of incorporating user centered design (UCD) tasks and the quick way to deal with incrementing the usability of a service or commodity. They have addressed the likenesses between the two cycles which empower integration. The two cycles depend on nominal design in advance and regular criticism from customers. Both are human-centered (Kristiadi, Udjaja et al. 2017).

The design process (Lowry 2019), (Google. Collaboration features 2020) followed to create the UI is presented below-

1. The process for the implementation model first starts with thorough research by surveying scientific papers and articles. This also involves surveying people using various different methods. There are numerous methods for identifying the sample space for research. College students were targeted for data collection. After random sampling over several runs, suitable youngsters were identified who voluntarily expressed willingness to participate in this

research. Apt knowledge of social media, awareness of various real time difficulties faced on a day-to-day basis were a prime motivation for the choice of these subjects for research. Initially, 60 of these students were identified, but due to factors uncovered on further examination: namely erroneous data received, lack of interest in participation, and lack of adequate background knowledge to allow participation, 30 respondents were selected to participate in the study. The sample space included 60% males and 40% females. As per (Kristiadi, Udjaja et al. 2017), proof of concept was implemented with these 30 respondents, and it is expected that this number will gradually increase in future experiments.

2. From the inputs collected during the above survey, extraction of suitable data has been performed with analytical tools. Additionally, Word Cloud was developed using R programming. This was needed in order to interpret what the users liked and what trends were observable. These tools removed the stop words thereby facilitating focus on words that provided insight. To help build the personas for the users an affinity diagram was developed wherein the survey acted as input for the questions that were asked, and the ideas generated served as possible solutions. These were written on sticky notes and stuck on the whiteboard. These were then arranged into groups, and appropriate new categories were formed in this brainstorming session. The personas thus built gave an idea of the users' characteristics that made them target users, thereby facilitating the identification of a user base. The brainstorming session also helped in distinguishing the features that were critical for the application besides providing prioritization of features.
3. The team members then thought of possible solutions to the problems that were perceived during the brainstorming session. This helped guide and constrain the designing and the prototyping phase.
4. After extensive survey, it was observed that one of the best platforms for understanding the basics of designing is Figma. The latter advocates a paradigm shift by shifting the mind from taking the design for granted, to understanding why a design is made the way it is. Besides the aesthetics, Figma also demonstrates building an intuitive mindset. Design platforms like Dribble were also used to create a community for asking design questions as well as carrying out research. After these tasks, the basic sketch of the application was designed and the modules identified. The flow of the product was decided and wireframes drawn. Subsequently, this led to the information architecture. Visual effects were added at the end and some extra further details.
5. The prototype thus delivered at the first iteration was tested for errors in user flow and refined to remove any errors. After sufficient brainstorming and idea development, the prototype was checked until it appeared to be error free.

These UCD and agile processes combined to deliver excellent results which satisfied customer requirements. Although these products needed to be tested after every iterative cycle, the tests offer some of the best insights as to how the user would want the product to be. A wide number of people will in general utilize products that are straightforward and function as expected. In the unique situation of this product design, system usability played a significant part in producing the apparent quality for its clients. Usability is the analysis of the crossing point between systems and clients' activities and desires in regard to use. Since numerous product items have been considered to be insufficient to address clients' needs, careful consideration has been given to ease of use. There are many definitions of usability that have been standardized since 1991 by several researchers such as Weichbroth (2020). Some of the definitions found in ISO/IEC 25010 are still currently in use which describe usability as the extent to which a system/product can be used by specific user in-order to attain specific goals along with features such as efficiency, effectiveness, satisfaction in a specified context of use (Fathauer and Rao 2019), (Almughram and Alyahya 2017).

The survey conducted in the current work is based on the model presented in (Goel and Goel 2016). The main purpose of the product is to be able to serve the intended audience in such a way that even users who are not experts in that field can utilize the product extensively. To incorporate this approach, numerous principles have to be appraised during the product's design process. Focusing on the UX part, various characteristics are to be considered since they can have huge impact, namely:

- Design of the product in accordance to user wants and desires or primary objectives.
- Product constraints and abilities.
- Product IA, aesthetics, and surface look.
- Product purpose.

For identification of user wants and needs TCSD is quite helpful for providing the initial steps. The steps are listed below: -

- Identification: finding out problems faced by the user regarding the chosen field, converting it into a task and providing a description for the problem.
- User-centered requirement analysis: deciding whether to keep or discard the outcome of the analysis.
- Design as scenario: where the design of the system and the data that is required by the system is presented in terms of a simulation.
- A final evaluation, called walkthrough evaluation to finalize the system design.

A questionnaire is prepared to evaluate the requirements and the type of components that are preferred. For scaling the requirements, the Likert scale has been used. This is used interchangeably with a rating scale. The Likert scale is a scale that is used to quantify the frame of mind, viewpoint and outlook of an individual or class of individuals. The responses and feedback to the questions will provide the degree of approval or disapproval for a certain component. A Likert scale can either have an odd or even number of options available. It usually has the format-

- Score 1: Strongly Disagree
- Score 2: Disagree
- Score 3: Neutral

- Score 4: Agree
- Score 5: Strongly Agree

Sometimes instead of 5 scales, an even scale of 4 or 6 is also used, forcing people not to select a neutral option. The latter has been used in the analysis of each component. So, this feedback will be involved in the stages of research carried out. The format we have used is-

- Score 1: Strongly Disagree
- Score 2: Disagree
- Score 3: Agree
- Score 4: Strongly Agree

6.6 EXPERIMENTAL RESULTS

Mockup and walkthrough evaluation is presented in this section.

6.6.1 Identification and Observation

6.6.1.1 Research and Identification of the Problem

People in daily life struggle at times to get jobs done due to lack of proper resources. This leads to long days of online searching and ineffective person to person questioning, leading to time wasting. This problem has been identified and a wide range of people are interviewed and, based on the survey a conclusion is reached. An approach is presented to them. A concept of a platform where they would be able to ask questions and get the answers directly and collectively is shared. It was asked whether the concept is helpful to them. Over 30 people were interviewed. We initiated the design process based on these responses.

6.6.1.2 User Identification and Purpose

The user's aim is to be able to get their job done within a few days. For this, a platform that has an easy-to-understand interface will help to lead a user through the product and post their query. Information will be provided by diverse groups of people and speedy completion of the task will follow. People having various daily tasks from diverse groups are the target audience.

6.6.1.3 Conversion into Objectives and Goals

The target audience's difficulties are taken into account and features that will fit into the UI are examined, guiding the user through the product along with assisting the process of collection of information.

6.6.1.4 Deduction of Variables

The components used in the interface will be sorted into variables deduced for testing purposes. In the test, UI dimensions are considered as shown in Table 6.3(a). Modules and descriptions of each module are provided in Table 6.5. Further, each variable will

TABLE 6.3
Various Tables Developed for Score Assignments

(a) Dimensions for Measuring

No.		Dimensions
1	User Interface	Typography
2		Contrast
3		Scale
4		Color
5	User Experience	Usefulness
6		Ease of Use

(b) Score assignment for UI

Score	Score Interpretation
1	Very Dissatisfied
2	Dissatisfied
3	Good
4	Great

(c) Score assignment for UX

Score	Score Interpretation
1	Very Useless/ Very Difficult
2	Useless/ Difficult
3	Neutral
4	Useful/ Easy
5	Very Useful/ Very Easy

be ranked by a number (1, 2, 3, and 4) in accordance with a feature. In testing the UX, there are two dimensions based on usability methods using the USE Questionnaire which will be given a variable that is shown in the table. Shown below in Tables 6.3 (a, b and c) respectively are dimensions for measuring, score assignment for UI, and score assignment for UX.

6.6.1.5 Questionnaire Design

6.6.1.5.1 Survey for Research

- Respondents were identified on the basis of their names, ages and occupations.
- They were asked questions on how difficult they found it to search for information on the internet, how long it took to get the required information and arrive at a decision, how frequently did they need the internet for searching information to get their job done and whether relatives, friends and neighbors were helpful in the process.
- Along with these, they were asked whether they would prefer a platform solely based on this, unlike social media apps, and whether this would help them.

6.6.1.5.2 Survey for Mockup

- After getting the mockup ready, following the wire-framing process, the same respondents were asked to give a review of the mock up UI.
- Respondents were told to give a response based on the prototype created for them. Based on the variables deduced, respondents were asked to rate the UI modules.

6.6.1.6 Questionnaire Data Processing Method

The questionnaire has no option for neutral scaling thus the respondents were forced to choose one of the poles. The questionnaire is based on Likert scale with positive questions. Based on the score assigned for the variables, statistics are calculated from the feedback received. The score of 1 represents the least value of that dimension and keeps on increasing up to 4 for scaling in UX. A score of 5 represents the maximum value of the variable in UI. The lowest value represents useless or very difficult UX or very dissatisfied when considering UI. The following is considered only for the second part of the survey.

$$\begin{aligned} &\textit{Total number of intervals} = \textit{numbe of options provided} = 5 \\ &\textit{Interval range} = 100 / \textit{Total no. of intervals} \\ &\qquad = 100/4 = 25 = \textit{Li}\,\text{ker}\,t \end{aligned} \tag{1}$$

Thus, scores allocated to the dimensions can be understood as per the following:

- Score 0% - 25% = Very Useless / Very Difficult / Very Dissatisfied
- Score 26% - 50% = Useless / Difficult / Dissatisfied / Rare
- Score 51% - 75% = Useful / Easy / Satisfied / Often
- Score 76% - 100% = Very Useful / Very Easy / Very Satisfied / Very Often

This means that the lowest score possible = Total number of respondents who chose alternative 1 x 25%

Similarly, the highest score possible = Total number of respondents who chose alternative 4 x 100%

Score interpretation = Total score calculated ÷ Total number of respondents

Thus, the total score will always range from 25% (lowest) to 100% (highest).

The rule for producing a corresponding evaluated value for each dimension and each page is:

$$\begin{aligned} X_1 &= \text{Score 1 value x the number of respondents who picked alternative 1} \\ X_2 &= \text{Score 2 value x the number of respondents who picked alternative 2} \\ X_3 &= \text{Score 3 value x the number of respondents who picked alternative 3} \\ X_4 &= \text{Score 4 value x the number of respondents who picked alternative 4} \end{aligned} \tag{2}$$

$$\text{Thus Total Score} = \sum X_i \tag{3}$$

For understanding the likability of a particular feature for a particular page:

$$\text{Total Score} \div \text{Total no. of respondents} = \sum X_i \div 22 \quad (3)$$

Finding Values of X and Y

- Y = highest score Likert x number of respondents (Scores 4) "Note Weight Value"
- X = Lowest score Likert x number of respondents (Lowest Score 1) "Note Weight Value"

$$\text{Index \% formula:} (\text{Total Score} \div Y) \times 100\% \quad (4)$$

6.6.1.7 Questionnaire Results

6.6.1.7.1 Survey Result from Research

The questions designed for the first half have been represented in pie charts, inferring a solid conclusion that regardless of the fact that it might be easier to search information on the internet and that it might take a day or two to reach a conclusion, people preferred to post their query on a platform which is made solely for the purpose of getting their jobs done as it might help them reduce time wastage. 22 people were interviewed. According to that, 8.8% people found it very difficult to search information from the net and 29.4% people said it is hard. Even though most people (38.2%) said it was easy to search information, people preferred that it would be more helpful if there was a platform to post their queries because people mostly rely on an internet search to get their job done instead of friends, relatives or neighbors. The time taken to arrive on a decision doesn't take much as mentioned by 23.5% or it takes a day or two for 44.1% of the people. Detailed result analysis from questionnaire data is illustrated in Figure 6.2

(a) Degree of difficulty to search information.
(b) Preference for relatives, friends, neighbors for information.
(c) Time taken for arriving at a decision.
(d) Preference for a query posting platform.

6.6.1.7.2 Survey Result from Mock Up

The results from the survey are provided in the ensuing tables. The results are based on the survey carried with 22 people and shown in Table 6.4

6.6.2 User-Centered Requirement Analysis

6.6.2.1 Story Board

The end user will engage with the product until he has achieved his goals. The requirements extracted from the user will form the goals for the product. The primary requirement of the user involves - posting, searching, and updating a task. The secondary goals or less important requirements are: being able to save the updates, viewing the reputation of the tasks or updates, saving the update received, downvotes

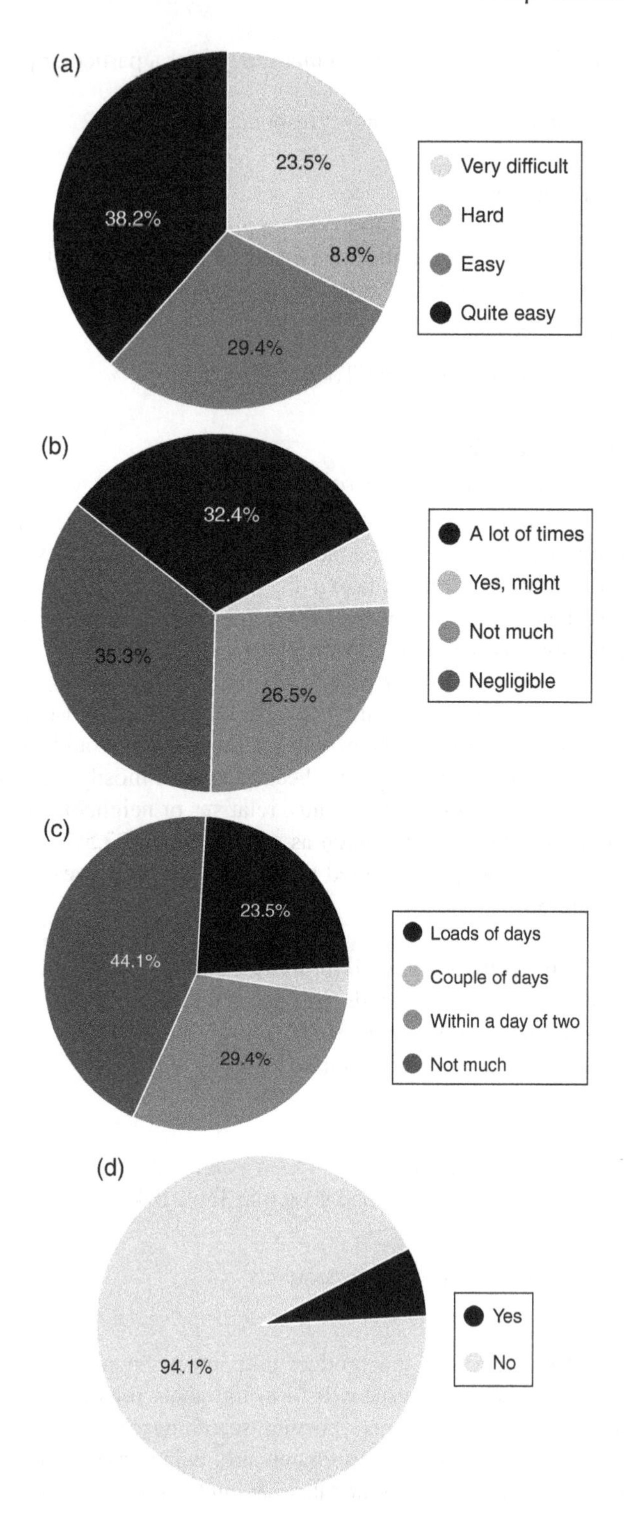

FIGURE 6.2 (a to d) Result analysis of questionnaire data.

TABLE 6.4
Average Results Based on Typography, Contrast, Scale and Color (in %)

		Respondents Options				Score Interpretation
Sr No.	**Pages**	**1**	**2**	**3**	**4**	**(%)**
1.	Splash Screen 1	6.65	22.82	23.91	46.74	77.74
2.	Splash Screen 2	6.52	22.83	28.26	42.39	76.63
3.	Log In	5.43	20.65	27.17	46.74	78.8
4.	Sign Up	6.52	20.65	25	47.83	78.53
5.	Verification	7.61	23.91	23.91	44.57	76.36
6.	Step 1: Personal Details	4.35	26.09	22.83	46.74	77.99
7.	Step 2: Adding Certificates	4.35	28.26	23.91	43.48	76.63
8.	Step 2: Profile Picture	4.35	25	28.26	42.39	77.17
9.	Step 3: Adding Friends	6.52	25	26.08	42.39	76.08
10.	Profile (Home)	4.35	23.91	23.91	47.83	78.80
11.	Search/Update	4.35	21.74	27.17	46.74	72.28
12.	Trending	5.43	20.65	27.17	46.74	78.80
13.	Notifications	4.35	21.74	25	48.91	79.62
14.	Post a Task	3.26	21.74	30.43	44.57	79.08
15.	Saved Updates	4.35	19.56	26.09	50	80.43
16.	Discover People	5.43	18.48	28.26	47.83	79.62

and the like. The user will have to log in or sign up to generate a profile on the platform before beginning to use the product. Other user profiles can also be checked. The achievement section shows partial verification of how qualified a person is. Mostly the main goals are provided to a user in the form of bottom navigation bar also referred to as the navigation bar.

6.6.2.2 Conceptual Model

This stage is based on the previous stage of the storyboard. The purpose of the conceptual model is to assist in the designing stage in the form of a paper prototype form-wireframe design. Table 6.5 presents the conceptual model to be used in developing the website, along with a module description.

The whole platform for the mobile app has been segregated into modules. There are introductory screens to give a brief idea of the application. Users have to log in or sign up if new. New users will be verified via CAPTCHA and OTP. A profile of the user is generated. One can discover more groups and people check their saved updates and delete the task or hide the task. A bottom navigation bar is provided for every page. The navigation bar provides easy access. The profile is the home page. One can directly go to the trending page and post or update a task. One can up-vote, down-vote a particular task or update. New notifications can be viewed by clicking the bell icon.

6.6.2.3 Wireframes

Figures 6.3 a– h illustrate in detail, the actual wireframes developed by us.

TABLE 6.5
Description of Modules

No	Page	Component	Details
1	Splash Screen 1 And Splash Screen 2	Status bar	Contains current time, network availability bars, WIFI signal strength, battery
		Heading	Contains a short introductory line
		Illustration	Contains a graphic describing the product as well as the introductory line
		Buttons	Contains "SKIP" and "NEXT"
2	Log In	Status bar	
		Overlay Layout	Contains a blank background to accommodate other components
		Input Fields	Contains two input field one for username and one for password, icons representing user and lock respectively, placeholder text
		Buttons	Contains log in button, Log in/ Sign up button panel
3	Sign-Up Page	Status bar	
		Overlay Layout	
		Input Fields	Contains input fields for name, gender, birthday, mobile number, recovery email address, username, password, text area for confirmation CAPTCHA, labels on top respectively, placeholder texts, drop downs for numbers and texts
		Buttons	Contains Log in/ Sign up button panel, "Save and Proceed", "Create" button
		Icons (Best 20 Example. YouTube. 2020)	Contains refresh, sound, help icons
4	Verification Page	Status bar	
		Overlay Layout	
		Illustration	Contains image alerting the reception of OTP
		Heading	Contains the text "Verification"
		Subtext	Contains title under verification heading and resend message
		Input Field	Contains blank dashes for the OTP number
		Buttons	Contains verify button
5	Verified Page (Delay)	Status bar	
		Overlay Layout	
		Heading	Contains the text "Verified!"
		SubHeading	Contains the text that the user has joined the platform
		Illustration	Contains image showing successful registration

TABLE 6.5 (Continued)
Description of Modules

No	Page	Component	Details
6	Profile Info Page (Step 1)	Status bar	
		Overlay Layout	
		Navigation bar	Contains all the three steps, personal information to be displayed on profile, uploading profile picture, finding people on the network
		Input Fields	Contains personal information like current education, work, internship, graduation details and current city and their labels
		Privacy Selector	A drop down for selecting degree of privacy on personal information, indicated by world icon
		Buttons	Contains Skip and Save & Continue buttons
7	Profile Picture Page (Step 2)	Status bar	
		Overlay Layout	
		Navigation bar	
		Uploading area	Contains area where the uploaded image will show and a plus sign where the number of current images uploaded will be shown
		Fixed text	Contains a text asking to upload certificates, achievements
		Buttons	Contains a button for upload ("Upload"), skipping the step and save & continue
8	Profile Picture Page (Uploading Achievements Step 2)	Status bar	
		Overlay Layout	
		Navigation bar	
		Uploading area	Contains area where the uploaded image of the user will show
		Fixed text	Profile picture
		Buttons	Contains a button for upload ("Upload Profile Picture"), skipping the step and save & continue
9	Find Friends	Status bar	
		Overlay Layout	
		Navigation bar	
		Profiles	Contains profile picture of the user and Profile name along with "Add" and "Remove" buttons to add or remove them from the network
		Buttons	Contains button to skip the process

(continued)

TABLE 6.5 (Continued)
Description of Modules

No	Page	Component	Details
10	Profile	Status bar	
		Overlay Layout	
		Profile Picture	Contains a space for user to add profile picture and a plus icon adding the picture
		Profile Heading	Contains Profile Name as well as Profile status and edit profile button
		Personal Information Section	Contains details about the user about their education along with their icons
		Achievement Section	Contains the text "Achievements" along with an icon to identify and information about the individual user's achievements in the form of images (mostly used for measuring the degree of a verified of user)
		Recent Activity	Contains the heading "Recent Activities" along with an icon to identify it, a subText, recently posted task in the formed network and people who have updated it
		Button	Contains profile editing in the profile header, hamburger icon button
		Bottom Status Bar	Contains icons showing home, search, trending and notification
11	Side Page	Status bar	
		Overlay Layout	
		Profile Picture	
		Profile Heading	Contains Profile Name as well as Profile status and see your profile option button
		Buttons	Contains features such as discover people, discover groups, saved updates, settings and privacy, help and support and log out
		Bottom Status Bar	
12	Post task/ Update task page	Status bar	
		Overlay Layouts	
		Profile Picture	
		Profile Heading	
		Posting Task boxes	Contains tasks posted previously by people along with their profile name, profile picture, location if enabled. Consists of a set of features

TABLE 6.5 (Continued)
Description of Modules

No	Page	Component	Details
		Posting Task boxes	Contains tasks posted previously by people along with their profile name, profile picture, location if enabled. Consists of a set of features
		Feature Set	The tasks and updates each contain common features of up-voting, down-voting, viewing, and forwarding represented by icons. Additionally, task contains a reply feature and update contains a save feature represented by icon
		Buttons	Hamburger Icon button
		Bottom Status Bar	
13	Post a task page	Status bar	
		Search Bar	Contains text field search for searching updates
		Task Area	Consists of a topic title and writing the task area.
		Input Fields	Contains drop downs, button selection and multiple feature selection to disable any feature for the people before posting any task
		Buttons	Buttons consisting post, upload images (represented with an icon), detect location (and corresponding message "activated" if set on) and hamburger icon button
		Bottom Status Bar	
14	Search/ Update by keyword page	Status bar	
		Search Bar	
		Trending Task	Contains list of all the tasks posted as well as newly posted task in the network with the profile picture and name and initial part of the task text
		Buttons	Hamburger icon button
		Bottom Status Bar	
15	Notifications Page	Status bar	
		Search Bar	
		Notification bars (Best 20 Example. YouTube. 2020)	Contains the profile name and picture that has upvoted, downvoted, viewed, forwarded, saved, replied to any of your tasks or updates
		Buttons	
		Bottom Status Bar	

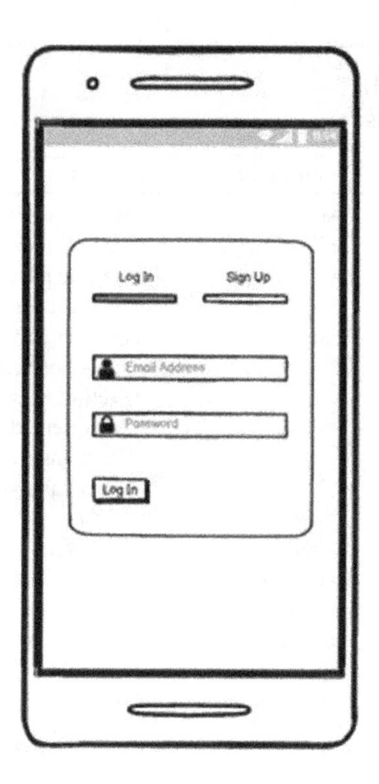

FIGURE 6.3(a) Wireframe of Log-In Page.

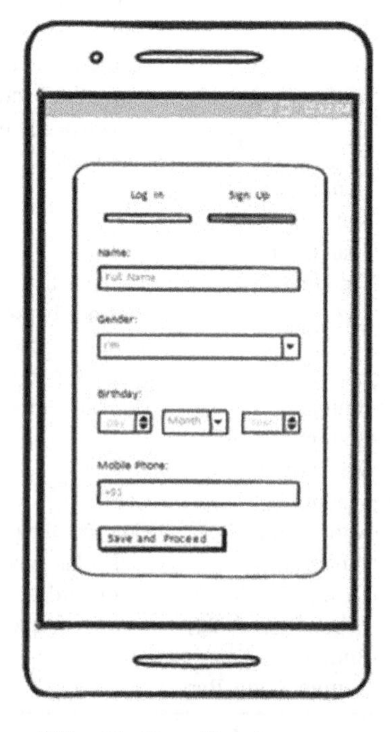

FIGURE 6.3(b) Wireframe of Sign-Up Page, Step 1.

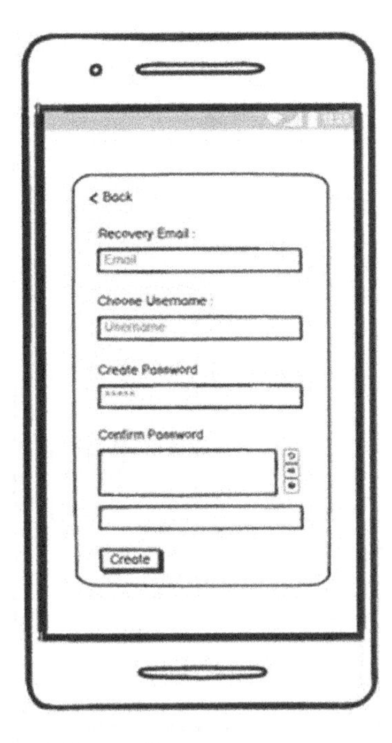

FIGURE 6.3(c) Wireframe of Sign-Up Page.

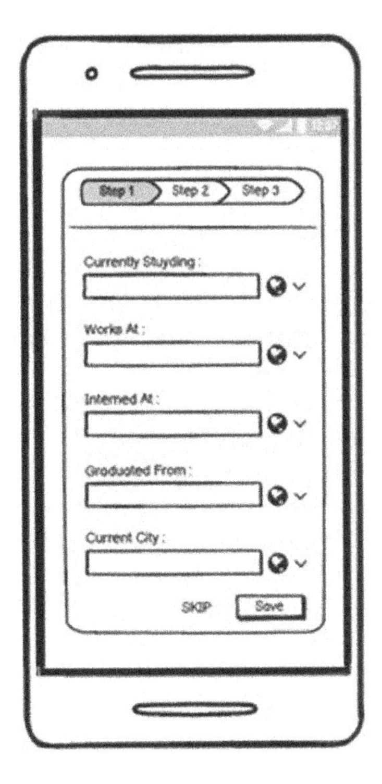

FIGURE 6.3(d) Wireframe of Step 1.

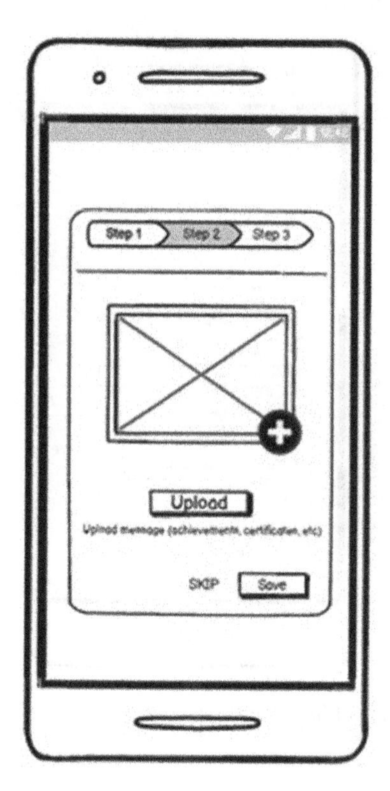

FIGURE 6.3(e) Wireframe of Step 2 (Achievements).

6.7 DESIGN AND ANALYSIS

6.7.1 Design

These are the first two introductory screens with illustrations on each. The illustrations are so designed as to be able to describe the product in brief. Effects are added for the 'NEXT' and 'SKIP' buttons. When clicked, it will be highlighted for a few milliseconds in the delay page. Figure 6.4 ('a' through 'd') illustrate the real screenshots developed by us.

A button panel is provided at top of the form. Smart Animate is used to show that on selection of Sign Up the panel slides toward it. On the Sign-Up Page necessary details along with drop downs for a few fields and mobile number is required for OTP verification at a later stage for registration of the user in the clickable prototype built.

The buttons are provided with a shadow of color #000000 with 25% visibility and 6 units of blur feature in 3 units x direction and 4 units y direction. The color of the buttons changes on clicking them. Shadow is provided to buttons to distinguish them as buttons, from the flat surface.

The main sign-up page is followed by a recovery email just in case the user forgets the username or password of the account. A text-based CAPTCHA is provided which is a type of challenge response test used in computing to determine whether or not

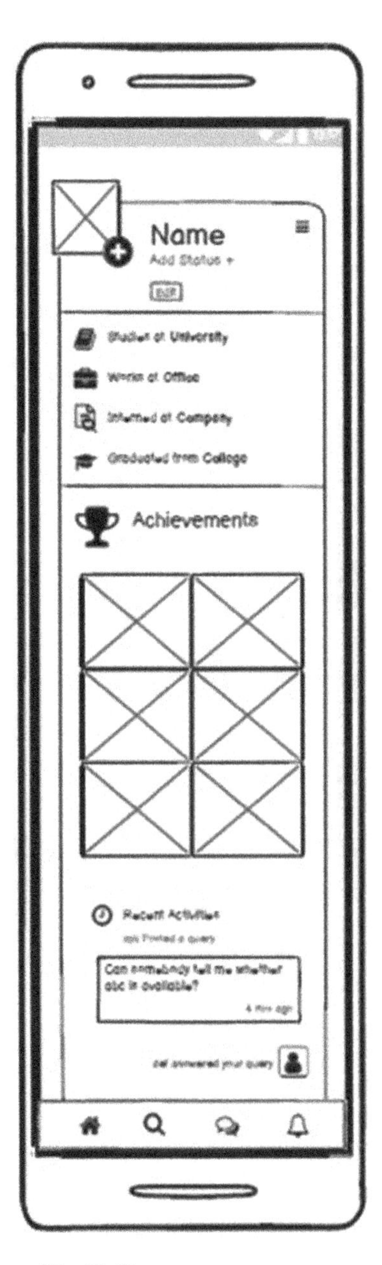

FIGURE 6.3(f) Wireframe of Profile Page.

the user is human. Many spammers bombard comment sections in an effort to drive a high click through rate to their websites and raise the search engine ranks. These comments are irrelevant to the post. This puts the app in danger. CAPTCHA helps you to avoid this by allowing only human beings to post comments. It is used to differentiate between human and automated access to websites. The test asks you to

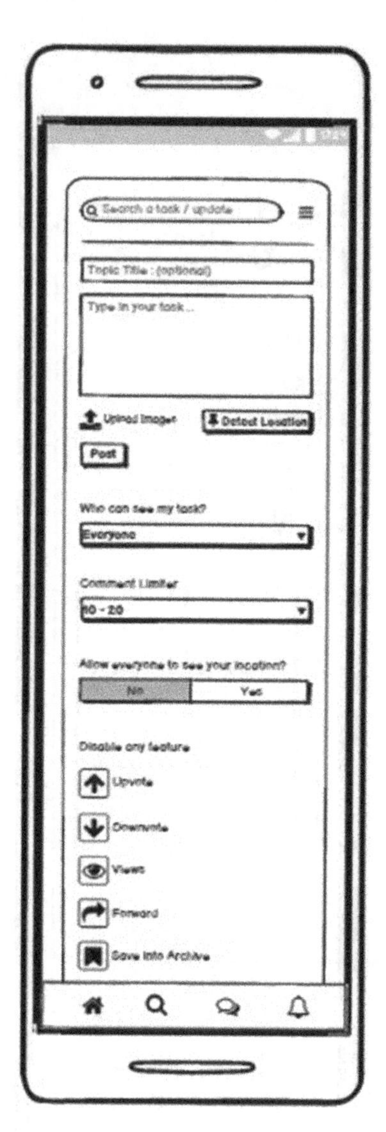

FIGURE 6.3(g) Wireframe of Posting a Task Page.

retype a disordered sequence of text. This type of test gives a great option for users who have a visual impairment and have trouble with other types of CAPTCHAS. The main drawback is that this CAPTCHA is easily solved by bots.

Buttons with icons like refresh, sound and help are provided on the right side. The refresh is meant to bring in a new CAPTCHA when the user cannot understand the CAPTCHA; the sound button will pronounce the CAPTCHA and the help button will provide further assistance if required.

After the sign up, the user is verified via OTP. The Figure 6.4(b) shows the screen for taking the input from the user. If the OTP hasn't arrived, then the 'resend' is

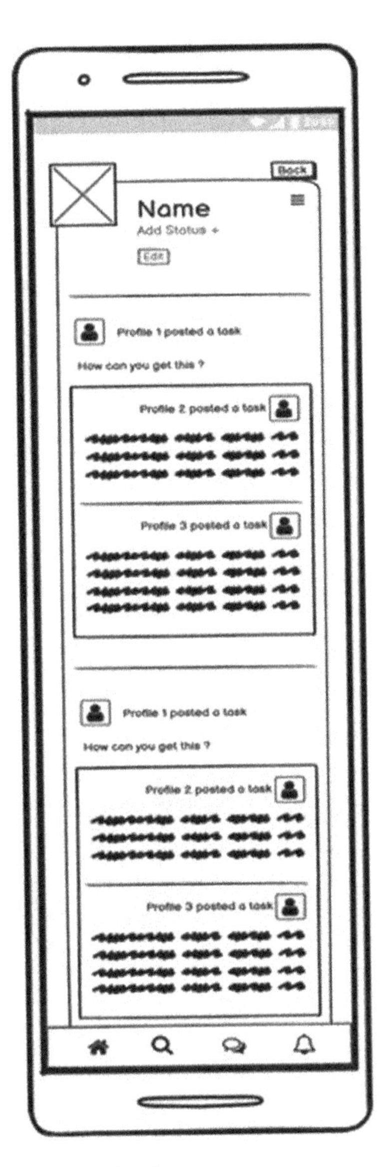

FIGURE 6.3(h) Wireframe of saved update.

FIGURE 6.3a–h Sample wireframes developed during Design phase.

clicked. This is present at the bottom of the screen. Once the user has been verified, a message is displayed to the user showing successful registration. The message after a few seconds slides to the personal details page.

There are three steps in the process of profile creation. Step 1 consists of filling out a form of personal details regarding education and other academically related fields. The user is required to fill in if it exists. All this information can be made private via the

FIGURE 6.4(a) Sign-Up Page.

FIGURE 6.4(b) Verification via OTP.

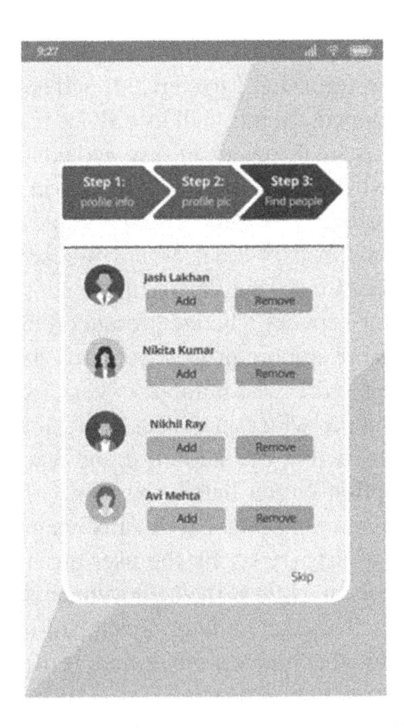

FIGURE 6.4(c) Step 3: Find people.

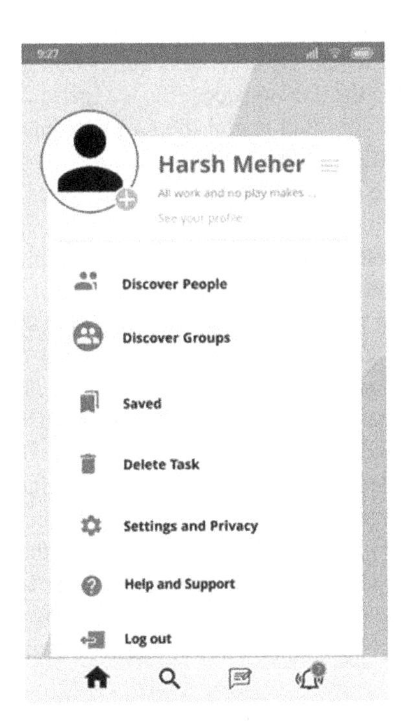

FIGURE 6.4(d) Side Page.

world icon drop down. It is a drop down to limit people from viewing this information. A button is provided to skip from step 1 to step 2. It will confirm first whether the user wants to skip the step. If skipped, the user will lose all the information if it was provided in those input fields. Otherwise, the user can save and continue. Step 2 is divided into two parts; first it will ask whether the user wants to upload any achievements and then it will ask the user to upload a profile picture. The profile generated is also the home page. This process can be skipped, or the user may proceed by providing information and clicking the 'save and continue' button. The last one is where the user gets to add or remove people from their network who are present on the platform currently.

Step 2 is considered more important as it requires the user to upload their life achievement and a profile picture. The number of certified documents uploaded can be spotted. However, to cater to wider audience, Step 2 may be skipped as well.

All the information provided in the steps 1, 2 and 3 will help to create a profile. The profile has an edit option button below the name of the user. Personal details are segregated from the remaining page. It is fully viewable by the user itself but depending on the degree of privacy set by the user during form filling, others may not be able to view this section. Next is the achievement section, where the user had to upload their certificates. This can be later developed into a prototype where this section can be used as a sign that the user is verified. If the user had skipped this step then the achievements section will be blank.

Each of the icons will lead to a different section. The home icon is for the profile, the search icon is for searching a task, write icon is for latest and trending tasks on the platform and the notification bell, which has the number of new notifications on top of the bell icon. Fixing of the navigation bar is quite common since a person does not have to scroll to the end of the feed to access the navigation bar. This helps in creating endless scrolling just like in other social media platforms. All the icons have been taken from the plug-in 'Material Design Icon'. It contains readymade icons in vector form. Vector networks are one of the most unique features in Figma. Most pen tools draw paths in a loop with a defined direction, always wanting to reconnect to their original point. Vector networks do not have a direction and can fork off in different directions without requiring a separate path object to be created.

The user can only view whatever is made public. All of these have been added later into the prototype. In the hamburger menu, saved updates can be viewed by clicking on the button 'Saved'. Corresponding tasks are also saved to save the users time in going back and reading the task. The user who posted the task and the updates saved by the user at the time are kept here along with the name of the people who updated it. These updates can be deleted when the user is done with the task. Once deleted, the remaining updates slide up. All the updates are kept in one rectangle so that the user can see all the updates collectively and separated from the task. Multiple tasks can be deleted by ticking the check box. Before deleting, the user will be asked whether they are sure that they want to delete them. On clicking 'Yes', it will delete. The selected ones will get highlighted, and a 'dustbin' icon will come up on selection of any one task.

Next in line comes the search button in the bottom navigation bar. The search page contains the search engine for the user to search a particular task by using keywords used in the task, hamburger menu and the list of people who have just posted a task.

The unread ones are highlighted in the color #FFF3F0 and the old ones appear in color #FEF9DE using some percentage of transparency. The list of people who have posted will appear with their profile pictures to the left side, along with their name in bold, to put emphasis on the user who posted, lastly with the task in light grey below their profile names and profile picture so as to attract the user to go and read it instead of leaving it incomplete. The notifications page is quite similar to the search page. Here the added difference is the action that another user has used along with its corresponding icon. The list contains people who have down-voted, up-voted or viewed your task or updates, updated your task, saved your updates or forwarded your tasks or updates.

6.7.2 Analysis

6.7.2.1 Analysis of Mockup Survey UI

Tables 6.6 (a, b, c and d) presents detailed analysis of UI and weightage per feature.

TABLE 6.6(a)
Detailed Analysis of UI

Sr No.	Properties	Average Score
1.	Typography	87.27
2.	Contrast	73.19
3.	Scale	77.72
4.	Color	74.93
Overall Average Score		78.28

6.7.2.2 Overall Satisfaction

TABLE 6.6(b)
Overall Satisfaction with the Product (in %)

		Respondents Option				
Sr No.	Properties	1	2	3	4	Average Score
1.	Satisfaction with the design of the product	0	4.2	29.2	66.7	90.7
2.	Satisfaction with the way it works	0	0	16.7	83.3	95.82
3.	Satisfaction with the placement of the icons	0	4.2	16.7	79.2	93.82
4.	Satisfaction with the features provided on the product	0	4.2	8.3	87.5	95.82
5.	I would recommend it to a friend	0	0	16.7	83.3	95.82
Average of all						94.40

6.7.2.3 Overall Satisfaction with Ease of Use and Usefulness

TABLE 6.6(c)
Ease of Use

Sr No.	Properties	Respondents Option				Average Score
		1	2	3	4	
1.	It is easy to use	0	0	25	75	93.75
2.	It is simple to use	0	0	20.8	79.2	94.80
3.	It is flexible	0	4.2	16.7	79.2	93.82
4.	It requires the fewest steps possible to accomplish what I want to do with it	0	0	16.7	83.3	95.82
5.	I don't notice any inconsistencies as I use it	0	0	29.2	70.8	92.70
Average of all						94.18

6.7.3 Observations from Table 6.6

The combination used in the typography section reflects the brand image perfectly. The fonts styles mainly used are Noto Serif and Khula which go in harmony. Minimum font styles have been used for clarity. The low overall average score is due to the color and contrast section. Contrast needs to be increased to make it more readable as well as the important information visible. As far as color is considered, it should tend more toward the cooler edge. Regardless of this, the customer satisfaction is high considering the way it functions, the ease of access to the buttons, icons and their corresponding meaning. These are a direct indication of factors like ease of use and usefulness. This indirectly indicates that people find it useful, not just in idea or concept but in terms of its function and interface.

6.7.4 Validating the Usability of this Survey

In this sub-section we present an overview of validation of current survey with the standard software usability metrics. Usability is a combination of the following measures - learnability, memorability, efficiency, errors and satisfaction (Usability. Google 2020), (Nielsen 2012), (Usefulness, Google 2015), (Kristiadi, Udjaja et al. 2017). Usability is the quality attribute that assesses how easy user interfaces are to use. It also refers to methods for improving ease of use during the design process.

- Following are some of the heuristics (Usability. Google 2020), (Nielsen 2012), (Usefulness, Google 2015), (Kristiadi, Udjaja et al. 2017). which fall under usability design, and considered in current work:
  1) Consistency and standards.
  2) Error prevention.

TABLE 6.6(d)
Usefulness

Sr No.	Properties	Respondents Option				Average Score
		1	2	3	4	
1.	It helps me be more productive	0	0	33.3	66.7	91.67
2.	It helps me be more effective	0	0	33.3	66.7	91.67
3.	It makes the things I want to accomplish easier to get done	0	0	29.2	70.8	92.7
4.	It saves me time when I use it	0	0	33.3	66.7	91.67
5.	It meets my needs	0	0	20.8	79.2	94.8
Average of all						92.50

3) Recognition rather than recall.
4) Aesthetic and minimalist design.

It is observed from current work, that consistency and aesthetics have a higher attainment level (over 90%) as compared to the others (over 70%). These are based on computations done on our end user questionnaires.

6.8 CHALLENGES

We envisage the following technical difficulties:

1. Creating numerous insightful cloud scenarios, validating them and ensuring these are error-free for further use.
2. As the next step, transforming the above system to configuration files.
3. The app needs to undergo thorough alpha, and beta testing. Further enhancements will be incorporated after considering the initial test results.

6.9 NOVELTY

- The very concept of developing a social app of this sort based on task and update approach is extremely innovative and highly useful to society. Use of this app will tremendously benefit the end users in terms of high precision in the answers and drastically reduced wastage of times. Also, due the insightful UI/UX, our app is successfully addressed at a wider segment of society.
- The characteristic features of our app are presented as under-
  - The screens are very intuitive.
  - It is not crowded so helps focus on the major elements.
  - Colors have been researched, and these colors are known to signal good sensations in the brain.
  - The first landing pages are the splash screens which inform the users how to use the application in brief descriptive pictures.

- The login page has very simple yet indicative icons that help guide the user easily.
- The sign up is not written in small text and actually has a different tab so that user is not confused by only log in.
- The sign-up page has systematic fields for information, with indicative icons that help user fill relevant user without much analysis.
- The verification page has big fonts and keypads for un-ambiguity and to lend focus to verification.
- The profile creation page adheres to common mechanisms. It has recent activities as well as edit button for editing profile.
- Facility for adding friends and uploading profile picture and achievements is provided.
- Hamburger menu provided in profile.
- Bottom navigation bar provided for direct access to main pages.
- The Q&A platform has been attested by users to be very easy to use.
- Trending page shows the latest tasks posted.
- Number of up-votes, down-votes, forwards, views, updates or saves can be seen in notifications.
- Tasks can be searched using keywords.
- Tasks can be closed once user's purpose is fulfilled.
- While posting a task, images can be uploaded, location detection can be done as well as restrictions on comment can be provided for privacy as well as disabling of features is also possible.

6.10 RESEARCH CONTRIBUTION

- Current work has been developed toward one of the projects at Hungreebee Technologies LLP (Hungreebee Technologies 2021) and aims to increase customer satisfaction and retention through the user friendly, intuitive and novel app.
- Also, an extensive survey has taken place pertaining to conceptual knowledge of UI/UX and related work by researchers in this field. Hence current work will be tremendously beneficial for budding researchers in the field of UI / UX and cloud-based app development
- We have extensively discussed the process of initiating design and subsequently creating a prototype. Analysis provided on the mock-up survey and the model of research can help other researchers in this domain to generate inferences that are helpful to start building their own UI and to come up with survey models suitable to their product prototype, respectively.

6.11 FUTURE WORK

- For storing the data of a large and varied audience, cloud services can be used as they provide security.
- To overcome the limitations of traditional load balancing (such as: significant effort for re-balancing, ineffective at coping with stragglers) we plan to

implement the Straggler Aware Execution (Role of Branding. Google 2020) for supporting analysis services in the cloud. This technique will foster increased parallelism.

- In current work, we have used Figma. However, in our extended work we propose to incorporate suitable graphical cloud-based modeling languages based on UML.
- For the Task-Update based scheduling in current work, we plan to incorporate a combination of QoS features to facilitate optimization in scheduling.
- Social media sites aid in the formation of connection among people on existing networks. These strangers connect on some particular basis like shared interests, common activities, common clubs, etc. Social media usually targets a wide range of people from all walks of life unlike some sites which might be for a group people speaking a common language, nationality-based individualities, religious or a diverse group (Figma. YouTube 2019). The users can further be categorized after implementation on a platform. The data collected from events occurring on the platform can further help and prioritize a particular section of audience.
- Moreover, trending topics which are of interest among a particular audience can be taken into consideration and based on the analysis and using that information, engagement of the user can be increased as well as interactivity.

Implementation of encryption algorithm AES (Kartit, Azougaghe et al. 2016, Castiglione, Choo et al. 2017) is proposed in future work. The trustworthiness and privacy of the information would thus be able to be guaranteed by giving access to the information just on fruitful verifications.

6.12 CONCLUSION

In this elaborate work we started with a rich conceptual background for developing a good UI/UX prototype. After rigorous survey, we noted the lack of products with the intent of providing people the platform to solve their daily issues. The color palette selection was done keeping the association of the color with particular features in line with human psychology. The decision for the color palette has been made after intensive research. The UCD system and USE was used to build the questionnaires for this project and responses were collected from more than 30 individuals. Lastly, some effects have been added in order to create endless scrolling, shadow effect for the buttons so to show that the button protrudes on the flat surface etc. These effects help user navigate the app and engage with the product.

We developed prototypes to help the users evaluate the flow along with the design. This has helped lead to the conclusion that there could still be improvements with respect to the user experience. From the analysis and user feedback improvements could be done on the layout, color and contrast to make it more user friendly and attractive on the right context. Based on ease of use and usefulness, the percentage of usefulness can be increased by making it friendlier to use and creating possible broad categories for people to select when posting a task, targeting only those users who are related to those categories. Improvements can be made for the trending page by making it simpler to use. The Proof of Concept of UI/UX work completed so far can be extended by deploying on cloud.

6.13 ABBREVIATIONS

- UI - User Interface
- ISO - International Standardization of Organization
- UI/UX Process - User Interface and User Experience
- IA - Information Architecture
- VD - Visual Design
- UE - User Equipment
- WCAG - Web Content Accessibility Guidelines
- CMYK - Cyan Magenta Yellow Black
- AAA - Authentication Authorization Accounting
- RGB - Red Green Blue
- TCSD - Task Centered System Design.
- AES - Advanced Encryption Standard
- RSA - Rivest–Shamir–Adleman
- SLA – Service Level Agreement
- UCD – User Centered Design

ACKNOWLEDGEMENTS

** The entire ownership of current work (concept, design and prototype) resides exclusively with the second author who had formulated the problem statement from his idea. Sachin Vahile has rendered all due technical support, guidance and co-operation toward the work, also reviewed the manuscript. Suja Panicker served as College Supervisor for the internship project which is described in this chapter. She coordinated and supervised the entire research paper writing process, guided the interns toward paper writing, contributed in content development, proofread and reviewed/revised the contents. Adrija Guin and Rahul Sethia served as Interns at Hungreebee Technologies LLP and implemented current work as part of their academic internship project with Sachin Vahile serving as Company Supervisor. Adrija Guin was responsible for creating the rough sketches, block diagrams, link between each screen, deciding on functionality as well as creating a prototype draft along with her proactive involvement in content development, proofing and reviewing the manuscript. Rahul Sethia was responsible for designing the website interface and taking decision toward deciding the symbols and icons to be used.

REFERENCES

Almughram and Alyahya. 2017. Coordination support for integrating user centered design in distributed agile projects. *IEEE 15th International Conference on Software Engineering Research, Management and Applications* (SERA). IEEE, London, UK.

Barua. 2019. Gestalt Principles: Secrets of Hacking Human Brain by Design. medium.com/ieee-sb-kuet/gestalt-principles-secrets-of-hacking-human-brain-by-design-85401fe6880d

Bernal, Cambronero, Valero, Núñez and Cañizares. 2019. A Framework for Modeling Cloud Infrastructures and User Interactions. *IEEE Access* 7:43269–43285.

Brien. 2010. The influence of hedonic and utilitarian motivations on user engagement: The case of online shopping experiences. Interacting with *Computers* 22(5):344–352.

Cartwright. 2020. The Designer's Guide to Color Theory, Color Wheels, and Color Schemes blog.hubspot.com/marketing/color-theory-design.

Castiglione, Choo, Nappi and Narducci. 2017. Biometrics in the Cloud: Challenges and Research Opportunities. *IEEE Cloud Computing* 4(4):12–17, doi: 10.1109/MCC.2017.3791012.

Fathauer and Rao. 2019. Accessibility in an educational software system: Experiences and Design Tips. IEEE Frontiers in Education Conference (FIE).

Figma Tips & Tricks - UI Designer's Superpower. 2019. Youtube. youtu.be/Vo0sEPqArRQ

Fu, Xiangqian. 2010. Mobile phone UI design principles in the design of human-machine interaction design. *11th IEEE International Conference on Computer-Aided Industrial Design & Conceptual Design, 1*, 697–701, doi: 10.1109/CAIDCD.2010.5681254.

Goel and Goel. 2016. Cloud computing based e-commerce model. *IEEE International Conference on Recent Trends in Electronics, Information & Communication Technology* (RTEICT), Bangalore. 27–30, doi: 10.1109/RTEICT.2016.7807775.

Google. 2015. Usefulness- The Usability Foundations. www.interaction-design.org/literature/article/usefulness-the-usability-foundations

Google. 12 Benefits of Cloud Computing. 2020, www.salesforce.com/products/platform/best-practices/benefits-of-cloud-computing

Google. Collaboration features. 2020. www.figma.com/collaboration/

Google. 2020. The Difference between UX and UI Design and Why it Matters www.scalablepath.com/blog/the-difference-between-ux-and-ui-design-and-why-it-matters/

Google. 5 Most Common Misconceptions about UI/UX Designing. 2020 content.techgig.com/ux-designing/articleshow/75048662.cms

Google. 2020. The role of branding in UI design. tubikstudio.medium.com/the-role-of-branding-in-ui-design-e6cc247b6b66

Google. 2020.Usability., www.interaction-design.org/literature/topics/usability

Hungreebee Technologies LLP. www.HungreeBee.com Accessed 30 January 2021

Joo. 2017. A Study on Understanding of UI and UX, and Understanding of Design According to User Interface Change. *International Journal of Applied Engineering Research* 12(20) 9931–9935.

Kartit, Zaid, Azougaghe, Ali, Idrissi, H. Kamal, El Marraki, M., Hedabou, M., Belkasmi, M., Kartit, A. 2016. Applying Encryption Algorithm for Data Security in Cloud Storage. In: Sabir E., Medromi H., Sadik M. (eds.), *Advances in Ubiquitous Networking. UNet 2015*. Lecture Notes in Electrical Engineering, vol. 366. Springer, Singapore. https://doi.org/10.1007/978-981-287-990-5_12.

Kelola, Andalan, Using Task System Centered Design (TCSD) Method. Fourth International Conference on Informatics and Computing.

Kim. Bringing Micro-Interaction & UI Animation to Life Through Developer-Designer Collaborations. 2017. medium.com/capital-one-tech/bringing-delightful-micro-interaction-and-ui-animation-to-life-through-developer-designer-3c409bc326f

Kopf. The Power of Figma as a Design Tool. 2020. www.toptal.com/designers/ui/figma-design-tool (accessed 30 January 2021)

Kristiadi, Dedy Prasetya, Udjaja, Yogi, Supangat, Budiman , Prameswara, Randy Yoga, Warnars, Harco Leslie Hendric Spits, Heryadi, Yaya, Kusakunniran, Worapan. 2017. The effect of UI, UX and GX on video games. IEEE International Conference on

Cybernetics and Computational Intelligence (CyberneticsCom), Phuket. 158–163, doi: 10.1109/CYBERNETICSCOM.2017.8311702.

Kumar, Kandaswamy and Deepa. 2016. Data privacy model for social media platforms. 6th ICT International Student Project Conference (ICT-ISPC).

Lowry, Thomas 2019. Component Architecture in Figma, www.figma.com/best-practices/component-architecture/.

Marketing Color Psychology: What Do Colors Mean and How Do They Affect Consumers?. 2019. Youtube. youtube/x0smq5ljlf4.

Neumorphism: Why it's all the Hype in UI Design. 2020. Google www.justinmind.com/blog/neumorphism-ui/.

Nielsen. Introduction to Usability. 2012., www.nngroup.com/articles/usability-101-introduction-to-usability/.

Nurgalieva, Leysan, Laconich, Juan José Jara, Baez, Marcos, Casati, Fabio, Marchese, Maurizio. 2019. A Systematic Literature Review of Research-Derived Touchscreen Design Guidelines for Older Adults. *IEEE Access* 7, 22035–22058.

Rufo. Design for the Dark Theme. 2019 medium.com/snapp-mobile/design-for-the-dark-theme-9a2185bbb1d5.

Soeegard, Mads. 2020. Dressing Up Your UI with Colors That Fit www.interaction-design.org/literature/article/dressing-up-your-ui-with-colors-that-fit.

Vaniukov, Slava. 2019. Colors in UI Design: A Guide for Creating the Perfect UI usabilitygeek.com/colors-in-ui-design-a-guide-for-creating-the-perfect-ui/.

Varshney, Shweta and Singh, Sarvpal. 2018. An Optimal Bi-Objective Particle Swarm Optimization Algorithm for Task Scheduling in Cloud Computing. *2nd International Conference on Trends in Electronics and Informatics* (ICOEI), Tirunelveli. 780–784, doi: 10.1109/ICOEI.2018.8553728.

Verma, Vijay. 2019. medium.com/zomato-technology/why-we-switched-to-figma-as-the-primary-design-tool-at-zomato-1aa8fa931b0a.

Weichbroth, Paweł. 2020. Usability of Mobile Applications: A Systematic Literature Study. *IEEE Access* 8: 55563–55577 doi: 10.1109/ACCESS.2020.2981892.

Youtube. 4 stages of UI Design → UI Design Basics. 2018 www.youtube.com/watch?v=_7LZ14xtfOc

Youtube. Steps in the design process. 2018. www.youtube.com/watch?v=oSz5KTdKW88

Youtube. The 2019 UI Design Crash Course for Beginners. www.youtube.com/watch?v=_Hp_dI0DzY4&t=1150s

Youtube. Best 20 Example UI/UX Design For Mobile App | UI/UX Animation Design. 2020. youtu.be/d6xn5uflUjg

Youtube. Logo Design and Animation. 2020. youtu.be/l8o3WOldZ3c

Yun, You-Dong, Lee, Chanhee, and Lim, Heui-Seok. 2016. Designing an Intelligent UI/UX System Based on the Cognitive Response for Smart Senior. *2nd International Conference on Science in Information Technology* (ICSITech).

Zhafirah Indira Paramarini Hardianto and Karmilasari. 2019. Analysis and Design of User Interface and User Experience (UI / UX) E-Commerce Website PT Pentasada.

Zhang, Yu, Liao, Xiaofei, Jin, Hai and Tan, Guang. 2017. SAE: Toward Efficient Cloud Data Analysis Service for Large-Scale Social Networks. *IEEE Transactions on Cloud Computing*. 5(3):563–575, doi: 10.1109/TCC.2015.2415810.

Zhao, Ming, Gao, Yongpeng and Liu, Chang. 2012. Research and Achievement of UI Patterns and Presentation Layer Framework. *Fourth International Conference on Computational Intelligence and Communication Networks*, Mathura, 2012, 870–874, doi: 10.1109/CICN.2012.175.

7 An Efficient AI-Based Hybrid Classification Model for Heart Disease Prediction

Vaishali Baviskar, Madhushi Verma, and Pradeep Chatterjee

CONTENTS

7.1 INTRODUCTION

Prediction of heart diseases from patient health data records of glucose levels, heart-rate-variations, temperature levels, blood pressure levels, and so forth, requires the design of many mutually dependent operations to work in tandem with high efficiency. These operations include but are not limited to:

- Data acquisition from the human body, wherein different parameters like heart rate pattern, blood pressure levels, glucose levels, triglyceride levels, and so forth, are extracted. These parameters are segregated into different lists in order to distinguish between them.

DOI: 10.1201/9781003138037-7

- The acquired data is given to a pre-processing block, wherein different denoising and signal enhancement operations are performed. These operations improve data reliability and usability for the next blocks.
- Upon pre-processing, the data is passed to a feature extraction unit. In this unit, different primary and secondary features are extracted. These features include autonomic balance, blood pressure (BP), gas exchange, gut, heart, and vascular tone. The feature extraction unit is responsible for making sure that different heart diseases have different identifying features. Features of one kind of heart-disease need to have minimum similarity with features of other heart diseases and features of one-heart disease need to have maximum similarity with the same heart disease features across different users.
- These features are given to a feature selection unit, wherein redundant or similar features are removed from the set, and only maximally variant features are kept in the dataset.
- The output of the feature selection unit is given to a classification engine. This classification engine is responsible for finding out differences between the feature sets to classify them into different heart disease types. The classification unit uses algorithms such as: neural networks (NNs), support vector machines (SVMs), recurrent neural networks (RNNs), and so forth (Miškovic and Vladislav 2014; Masabo et al. 2019).

A significant proportion of research in heart disease detection is directed towards feature extraction and classification blocks. In this chapter, a brief study of different classification methods takes place, followed by the design of a proposed RNN based on a machine learning classification model. When combined, the classification methods are termed intelligent machine learning (ML) methods because they perform the classification of different heart diseases using previously trained data. The next section demonstrates different AI methods for the classification of heart diseases, and this will allow readers to identify best practices followed during classifier selection and to adopt them in their respective systems. This section is followed by the proposed RNN-based classification model and its performance evaluation. This chapter finally concludes with some acute observations about the proposed work and ways to improve the same (Fatma and Menaouer 2020; Kumar and Verma 2005).

7.2 RELATED WORK

Many researchers have worked and contributed novel AI-based techniques for heart disease prediction. Several ML and Deep Learning (DL) models have also been proposed to improve the accuracy of heart disease prediction.

7.2.1 Machine Learning Classifiers Used for Heart Disease Prediction

The heart is the central organ in human beings. Prediction and diagnosis of heart-related diseases need high accuracy, precision, and perfection. A slight fault can effect severe problems and possibly the death of an individual. Many people have died due to heart disease and an increase in the numbers has been observed. To tackle this situation, an essential requirement is accurate prediction to alert the patient about the disease well in time. ML is one of the practical and effective technologies designed for the testing. The technologies are built by training the models and testing them using relevant datasets. ML is a subdivision of AI, which includes a wide area of knowledge of machines that imitate human skills and capabilities. Considered from another angle, machine intelligence is enlisted to train the machine and help it learn to process the data by utilizing individual and hybrid models. The ML concept is to learn and acquire parameters from the normal profiles, and to use biological/genetic parameters as testing data, for example, sex, age, blood pressure, and cholesterol. This shows an good level of accuracy in comparison to the various parameters of different algorithms (Haq 2018, 2).

ML provides prominent support in AI to predict some types of heart attack event based on patients' everyday data. If doctors can predict heart disease accurately in advance, then it would help patients to avoid further consequences. ML is an efficient methodology that consists of training phases and testing phases, applying various types of algorithms to service the necessary need to determine the presence or absence of heart disease in a patient.

There are two kinds of ML algorithms given in Figure 7.1, that is, supervised learning and unsupervised learning. For heart disease prediction, both supervised and unsupervised algorithms are used by most of the researchers:

7.2.1.1 Supervised Learning

Supervised learning is the ML task of learning a function based on input-output pairs. It maps and assigns inputs and outputs to each other. It determines a function from labelled training data comprising of a set of training instances. The main aim of supervised learning is to predict the output by training it on a new data set. It requires

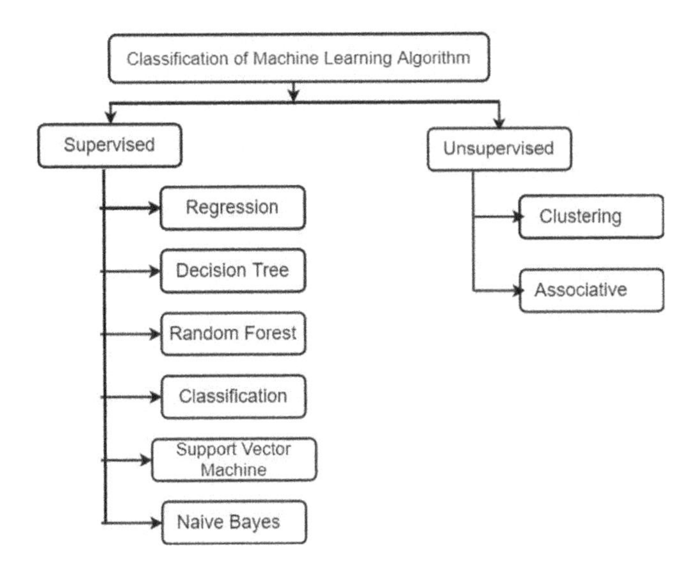

FIGURE 7.1 Machine learning algorithms used for heart disease prediction.

supervision and observation in order to train the model. It is categorized into regression, support vector machine (SVM), decision tree (DT), random forest, Naïve Bayes, and classification algorithms. This model produces an exact result. In supervised learning, the model is trained using the given dataset and only then an accurate output can be predicted. It comprises of algorithms like DT, linear regression (LR), SVM, Bayesian logic, multi-class classification, and so forth. As mentioned earlier, classification methods are broadly used in healthcare, as these methods and algorithms can process a very large data set. This data set can consist of numeric, textual, and image data. The commonly used classification methods in healthcare are: artificial neural networks (ANN), SVM, DT, nearest neighbor, and Naïve Bayes. ML with classification techniques is used for complex measurements in the medical domain and its applications. Recent classification methods have delivered more effective and intelligent predictions for heart diseases. Many studies have shown that combining algorithms with feature selection techniques and other algorithms to generate hybrid models have contributed towards correct prediction by using the optimum feature set.

7.2.1.2 Unsupervised Learning

Unsupervised learning methods use unlabeled data. It is defined as the process of learning without having any guide or teacher. It automatically and repeatedly works on the given dataset, discovers the relationships and patterns, and as per the generated patterns, it finds and classifies the relationships when a new dataset is assigned. It is constructed on the "self-sufficient" model concept. Initially, only the input dataset is delivered to the model in unsupervised learning. The main aim of unsupervised learning is to explore hidden patterns in an unknown dataset. It does not require supervision and observation for training the model. Unsupervised learning is categorized into association, and clustering types. As it trains, it learns day by day, by its own experiences like a small child, and as compared to supervised learning, this model can give less accurate outputs. It consists of many algorithms and methods like clustering, KNN, and the Apriori algorithm (Miao and Miao 2018).

Table 7.1 shows various supervised and unsupervised algorithm evaluations for 70% training and 30% testing in terms of accuracy when applied on the popularly used heart disease parameters dataset, that is, the Cleveland dataset from the UCI repository. For the implementation of Python programming, the Anaconda (Spyder) distribution is the best tool that can be used as it includes many header files and types of libraries that leads to correct predictions, making the work more precise and accurate.

7.2.2 Deep Learning Classifiers Used for Heart Disease Prediction

Prediction for any patient of the risk of heart disease is made using deep learning algorithms. Whenever the massive amount of patient information with high dimensional data is entered as a dataset, DL is applied, and an accurate prediction is generated. In deep learning, a model is trained directly from the images and text. For a vast labelled dataset, a neural network architecture with many layers is used (Bashir et al. 2017). In medical research, DL requires few processing steps because it passes through various stages of normalization and filtering. We have

TABLE 7.1
ML and DL Classifiers on Cleveland Dataset for Heart Disease Prediction

ML and DL classifiers for heart disease prediction	Accuracy	Precision	Recall	F-measure
SVM (linear)	85.71	84.09	86.04	85.05
Naïve Bayes	78.02	76.74	76.74	76.74
Decision Tree	79.12	77.27	79.06	78.16
Random Forest	81.31	82.5	76.74	79.51
Logistic Regression	82.41	80.0	83.72	81.81
K- nearest neighbour	80.21	83.33	80.0	81.63
XG Boost	82.41	81.39	81.39	81.39
Multilayer perceptron	72.52	70.45	72.09	71.26
Deep Neural network (200 epochs)	80.21	83.33	80.0	81.63

implemented a DL hybrid prediction model for accurate prediction of a patient who is expected to suffer from heart disease. We have proposed a combination of RNN and LSTM as a risk prediction for numerical and categorical heart disease. The accuracy achieved in the developed hybrid model is 96% (Miao and Miao 2018; Baviskar and Verma 2021).

Several DL techniques based on ANN, convolutional neural networks (CNN) and RNN were used by various researchers to deal with the accurate prediction problem for heart diseases (Kopiec and Martyna 2011; Kumar and Inbarani 2015; Xiao et al. 2020; Mohan, Thirumalai and Srivastava 2019).

CNNs are mostly used to order ECG images with various graphical data features. RNN and LSTM are progressive, forward, and backward thinking neural networks built with fixed-size input and output vectors.

Haq et al. (2018, 19) have used seven well-known machine learning algorithms, cross-validation methods, three feature selection algorithms, and evaluation metrics for classifiers' performance in areas such as: accuracy, sensitivity, specificity, execution time, and Matthews' correlation coefficient. All the classifiers checked for an execution time and accuracy with feature selection (FS) algorithms such as LASSO with k–fold cross minimum redundancy maximum relevance (mRMR) on selected features. Researchers have designed an intelligent system to classify healthy people and people with heart disease.

Miao and Miao (2018) proposed a deep neural network (DNN) classifier and diagnosis tool as a training model for accurate heart disease prediction. Applying the classifier as enhanced DNN, they achieved 83.67% accuracy.

Ashraf, Rizvi and Sharma (2019, 52) proposed a DNN technique to create an automatic heart attack prediction system. ML techniques were tested on multiple datasets for maximum accuracy. The proposed method introduced an automated preprocessing approach in data and removed the anomalies from the system.

Sowri et al. (2019) used DNN for prognosis prediction to predict high risk. The model used multiple RNNs to diagnose and learn. The proposed method achieved good accuracy.

Yang et al. (2020) used various techniques to build and predict a model. An automated health parameter recording device was used for continuous follow-up. They provided a three-year risk assessment prediction model based on a large population for cardio vascular disease (CVD).

Sharma and Parmar (2020) proposed the Talos hyper-parameter optimization model to predict cardiac risk. Classification was performed using Naïve Bayes, SVM, and Random Forest. UCI repository data was used to show that Talos hyperparameter optimization performed better than other classification algorithms.

Baccouche et al. (2020) proposed a framework for an ensemble-learning framework built on a bidirectional and unidirectional long-term memory (BiLSTM) model with a CNN that had an accuracy of 91% for the prediction of heart disease. A data pre-processing option with feature selection was implemented to increase the performance of the classifier. Heart disease parameter predictions using various ML and DL algorithms have been presented in the following Table 7.2.

7.2.3 Hybrid Models in AI

A hybrid model is an intelligent system that consists of a combination of techniques and methods in parallel from AI subfields such as: evolutionary methods, reasoning methods, supervised learning, reinforcement learning, and neuro-fuzzy systems.

TABLE 7.2
Survey of Heart Disease Prediction Using ML and DL Algorithms

Sr. no.	Author	Year	Dataset Used	Algorithm used	Accuracy
1	Amin Ul Haq et al. (Haq et al. 2018)	2018	UCI repository dataset	LR with cross-validation selected by Feature Selection algorithm Relief	89 %
2	Kathleen H. et al. (Miao and Miao 2018)	2018	UCI repository dataset	Enhanced DNN	83.67 %
3	Mohd. Ashraf et al. (Ashraf, Rizvi and Sharma 2019, 53)	2019	UCI repository dataset	Deep neural network	87.64
4	N. Sowri Raja Pillai et al. (Sowri et al. 2019)	2019	Dataset from patients	RNN with Genetic algorithm	92%
5	Li Yang et al. (Yang et al. 2020)	2020	Dataset from patients	Random Forest	78.7%
6	Sumit Sharma, Mahesh Parmar (Sharma and Parmar 2020)	2020	Heart disease dataset from Kaggle	Talos hyperparameter optimization algorithm (hybrid)	90.78%
7	Asma Baccouche et al. (Baccouche et al. 2020)	2020	Dataset from patients	BiLSTM and CNN-ensemble learning classifier	91%

To build a hybrid model, numerous methods are generally joined in a 2-stage manner, in which the primary phase is built on each classification or on clustering methods, which are used to pre-process the dataset. The outcome of the first stage, in other words, the processed data, is recycled to construct the following classifier as the prediction model. Especially, logistic regression, neural networks, and DT have been reused as the classification methods and k-means for self-organizing maps have been used to construct various hybrid models as clustering techniques.

7.2.3.1 Hybrid Classification Models

Hybrid classification models usually combine multiple incremental learning classification models, such that the inefficiency of one model is covered by one or more of the other models. For instance, decision trees are combined with Naïve Bayes, thus forming the NB tree classifier. The Naïve Bayes algorithm is suitable for targeting small range sequences, while the decision tree algorithm works better with long-ranged sequences. Thus, a combination of the two works on both long-ranged and short-ranged sequences are alike. Such combinations help build high speed, high accuracy, low complexity models, which may be applied to various datasets. Table 7.3 indicates different kinds of hybrid classification models and their potential application areas. By referring to this table, readers can get a reasonably good approximation of which model combinations must be used for which kind of application.

7.2.3.2 RNN Model Classifier in AI

A recurrent neural network (RNN) exists as a feedback inspired neural network, wherein the hidden layers are so connected that they recurrently learn from each other. A sample of the architecture of RNN can be observed from Figure 7.2, wherein the input layer data is given to the first hidden layer. The hidden layer neurons are so designed that they can accept data from the input and internal neurons.

Once the first hidden layer processes the input data, computed weights are given to the second hidden layer. The mentioned hidden layer performs further computations and optimizes these weights. The new optimized weights are recurrently given back to the first layer, where based on the feedback obtained, these weights are recurrently changed. Equation 1 represents the final weight equation of the neural network.

$$w_{(i+1)}=f(w_i, w_r) \tag{1}$$

Where $w_{(i+1)}$ is the new weight from the first hidden layer; f is the activation function for the network, w_i is the current weight of the layer, and w_r is the weight obtained from the recurrent layer. An RNN must have at least two layers. As the number of layers increases, there is an exponential increase in the network's computational complications.

Due to the rise in this computational complexity, there is a need to store substantial training weights. This is referred to as an exploding problem for RNNs. Moreover, there are other issues with RNNs, like vanishing gradient, wherein if large training sequences of similar weights are used, then the learning is stalled. Moreover, if activation functions like tanh or rectified linear unit (ReLU) are used, long sequences

TABLE 7.3
A Brief Comparison of Different Hybrid Classification Systems

Classification model	Details	Typical Application	Approximate accuracy (%)
Functional Trees (FT) (Miškovic and Vladislav. 2014)	Generate multi-variable trees, and each tree can be either Random Forest, Decision Tree, or any other tree structure which can be used for classification	DNA Sequences, heart disease cardio beat classification prediction	90
Multi Decision tree (Kumar et al. 2020)	Multiple decision trees are combined in order to segregate & classify clustered data points. Each cluster consists of similar data points so that the classification process can be designed to be more accurate	Land use land cover, Heart disease classification	91
NFE (Masabo et al. 2019)	Different kinds of feature extraction units are combined in order to discover the fine features for classification. These features are selected at random in order to get the best combination of selected features that gives the highest accuracy	Malware classification, human health prediction	93
MOEFC (Abdeldjouad et al. 2020)	Multi-Objective Evolutionary Fuzzy Classifier combines the fuzzification process with an evolutionary classifier, and the output classes are evaluated using defuzzification. The number of parameter classes depicts the number of objectives of this classifier	Heart disease prediction	79
AdaBoost (Abdeldjouad et al. 2020)	Adaptive Boosting usually combines multiple feature extraction units and boosts their weights in order to evaluate the best features for a given classification problem	Heart disease prediction	80
GFS-Logit Boost-C (Abdeldjouad et al. 2020)	Genetic Fuzzy System LogitBoost combines inputs with a genetic algorithm that uses AdaBoost like algorithms for feature evaluation. These are generally the most used classification systems with minimum overheads and high accuracy.	Heart disease prediction	94

TABLE 7.3 (Continued)
A Brief Comparison of Different Hybrid Classification Systems

Classification model	Details	Typical Application	Approximate accuracy (%)
FH-GBML-C (Abdeldjouad et al. 2020)	Fuzzy Hybrid Genetic Based Machine Learning is similar to Genetic Fuzzy System Logit Boost but does not have an AdaBoost, which reduces the classification accuracy of the system	Heart disease prediction	83
FURIA-C (Abdeldjouad et al. 2020)	Fuzzy Unordered Rule Induction uses a rule induction mechanism to orchestrate better classification rule selection for fuzzy inputs. It is a simple cascade classifier, thus has limited performance	Heart disease prediction	82
LRDA-GNN (Zhang, Kuma and Verma 2005)	It combines Logistic regression with Discriminant analysis to extract features and then uses a Genetic neural network for final classification. This results in a highly accurate classification system	Mammo-gram classification	93
MLP-PSO (Bouaziz and Boutana 2019, 149)	Combines a multi-layered perceptron based neural network with an improved particle swarm optimization algorithm to classify different kinds of signals. It has high accuracy and low chances of error due to the use of error reduction PSO in the system	Heart disease Cardio Beat classification	98
DGEC (Pławiak and Acharya 2020)	The deep genetic ensemble of classifiers is a series of connected classifiers, which are cascaded such that the outcome is a set of classifiers that learn from each other. Usually, multiple support vector machines KNN, SVM, and Radial Basis Feed Forward Neural Networks (RBFNN), and probabilistic NN (PNN) are used for forming a deep genetic ensemble network	Arrhythmia detection	99

(continued)

TABLE 7.3 (Continued)
A Brief Comparison of Different Hybrid Classification Systems

Classification model	Details	Typical Application	Approximate accuracy (%)
GABC (Muthuvel and Alexander 2019)	A combination of the Genetic Algorithm with Bee Colony Optimization is done to form a highly accurate classifier. Here, GA is used for feature selection purposes, while BCO is used for the final classification	Heart disease Cardio Beat classification	93
PSO-SVM (Kopiec and Martyna 2011)	A combination of PSO for feature evaluation and SVM for final classification is done for obtaining a highly sophisticated classifier	Heart disease Cardio Beat classification	94
Bijective soft set (Kumar and Inbarani 2015)	Combines the results of set theory with soft computing approaches in order to improve the classification performance of any signal processing system	Heart disease cardio beat classification Cardiac Arrhythmias classification	98
Hybrid Transfer Learning (Kudva and Guruvare 2020)	This network combines learning from different convolutional neural network architectures like AlexNet, GoogleNet, ResNet, VGG Net, etc., in order to evaluate the best features for classification of any signals. The signals can be images, audio, sensory signals, etc., which vary in the slightest form between each class	Uterine Cervix Images for Cervical Cancer Screening	91
HCFC (Xiao et al. 2020)	Combines hybrid classification framework with clustering in order first to recognize the best features for classification, and then use these features in order to classify the input data with the help of stacked, cascaded, or any other hybrid classification architecture	ECG, Stock Prediction	94
NB-SVM (Ingole, Bhoir and Vidhate 2018)	This system combines Naïve Bayes with SVM in order to first evaluate the best features and later classify these features into different classes	Text Classification	94

TABLE 7.3 (Continued)
A Brief Comparison of Different Hybrid Classification Systems

Classification model	Details	Typical Application	Approximate accuracy (%)
CBA-ANN-SVM (Torabi et al.)	A feed-forward neural network follows the clustering of feature points of data, wherein the final classification layer is replaced with a support vector machine classifier. The final classifier is similar to CNN and produces high accuracy	Forecasting Short-Term Energy Consumption	98
Feature weighting with hybrid classification (Asghar, et al 2020, 3484)	Each of the input features is weighted according to their importance. These weights are given to one of each classifier to discover the best classification model for the given input data	Spam detection	96
HRFLM (Hybrid random forest linear model) (Mohan,Thirumalai and Srivastava 2019,81545)	UCI repository dataset	Heart disease parameters classification	88.67

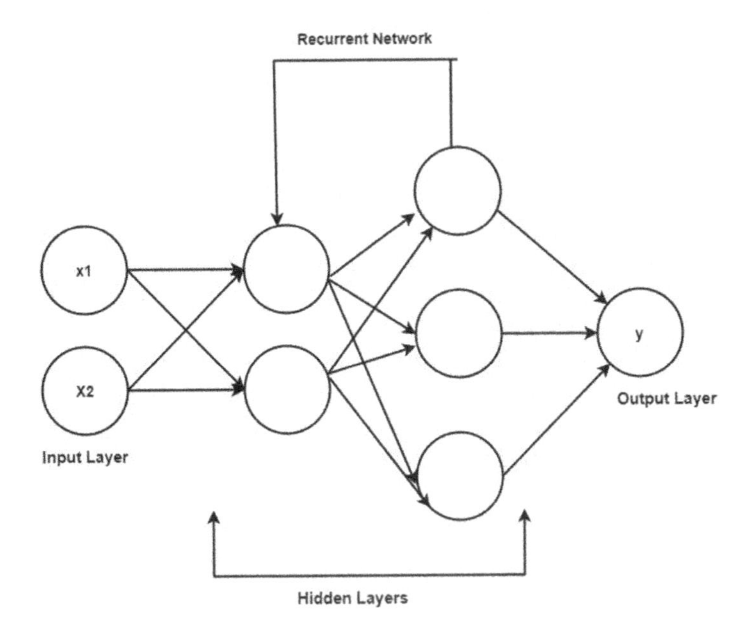

FIGURE 7.2 A sample RNN classification model.

cannot be processed. Even if other activation functions process long sequences, these weights are updated very frequently. Therefore, long short-term memory (LSTM) networks are generally combined with RNNs for better performance. Such networks are called hybrid networks and can be used for effective electrocardiogram (ECG) classification. This chapter introduces different kinds of hybrid classification networks and then takes the example of a very well-known Cleveland dataset for predicting heart diseases. Finally, this chapter concludes with a case study on a dataset with hybrid classification architecture in order to validate its efficiency.

The given table clearly shows that, deep learning classifiers that include the use of convolutional neural networks and clustering are best suited for multiple kinds of applications. Combining RNNs with LSTMs can be a good approximation for classifying heart disease-related parameters involved in the heart disease prediction problem.

A survey of various hybrid models used for heart disease prediction has been presented here.

Mohan, Thirumalai and Srivastava (2019, 81545) proposed a hybrid machine learning technique for an effective prediction of heart disease. They implemented a new method that finds major features to increase the accuracy of cardiovascular prediction by applying ML techniques. The prediction model is acquainted with various features combined with many well-known classification methods. ML methods were used to process and develop a raw dataset, which provided a new solution for the heart disease prediction problem.

Tarawneh and Embarak (2019) produced a prediction for heart disease comprised of a variety of prediction and classification methods. Researchers proposed a hybrid model approach that combined all methods and functions into one for accurate diagnosis and prediction. The proposed technique contains three stages: the initial phase is a pre-processing stage where the record is classified and filtered prior to any processing technique. The initial output is subject to numerous classification methods and evaluated to remove low-level performance. It then combines the outcome with the patient's present and past records to give a result and prediction about heart disease, whether positive or negative.

Xiao et al. (2017) proposed a hybrid model, which achieved noticeable results with performance metrics such as sensitivity, a confusion matrix, specificity, ROC, and accuracy. The hybrid classification model consists of the relief and rough set (RFRS) method, which handles both redundant and relevant features. An ensemble classifier with two subsystems, that is RFRS, features selection and classification. A 2- stage hybrid modelling was used, integrating RFRS, that is, RF with RS. The relief algorithm is an FS algorithm that chooses essential and relevant features by obtaining feature weights. It includes three stages: feature selection of data discrimination, feature extraction using the relief algorithm, and feature reduction using heuristic rough set reduction algorithm. For classification, C4.5, an ensemble classifier, was implemented. An accuracy of 92.59% was achieved with the cross-validation method.

Nikookar and Naderi (2018) proposed a powerful tool which performed better than other prediction models for heart disease and used reliable and ensemble hybrid models to assist the medical physicians in predicting and detecting heart disease. It evaluated the accuracy as 96%, the specificity as 93%, and sensitivity of 80%, which was achieved by applying the proposed hybrid ensemble model on the dataset of 278 instances. Here, researchers studied and investigated a hybrid model application and that performed better than other ensemble classifiers and has been proven to attain better performance. The performance of five widely used hybrid ensemble classification models was reviewed, including, k-NN, Naïve Bayes, SVM, random tree, and Bayes net where these basic classifiers are aggregated and the results forwarded to a new fusion classifier, such as: LogitBoost, AdaBoost, Random Forest and MLP for the diagnosis and prediction of the disorders of the heart has been evaluated.

Singh and Jindal (2018) researchers built a model for giving highly accurate predictions of heart disease. It was a combined approach of hybrid Naïve Bayes and the genetic method. The hybrid model of both these methods was referred to as hybrid genetic naïve Bayes model to predict higher accuracy for an implementation. They had used various algorithms like genetic algorithm and naïve Bayes. The programming platform used was Python 3.6. Data processing and classification technique was developed for prediction, as supervised learning. The performance parameters were accuracy, recall, and precision.

Agarwal and Ameta (2019) exhibit the work in two stages to increase prediction accuracy regarding cardiac issues. A new technique was proposed for the heart disease-related parameters like age, heart rate, cholesterol, pulse rate, and so on. In stage one, a new method was proposed in which parameters like cholesterol, pulse rate, and so forth were included alongside the age of a patient. This, in comparison to the previous studies, where only the age factor was considered as a primary attribute for prediction. In stage two, a new and efficient hybrid classification model was designed, which combined two different classification methods, in other words, KNN and SVM. The SVM extracted the dataset's significant features, and KNN acted as a classifier to generate the results. The proposed method's performance was better in comparison to other methods in terms of execution time and accuracy.

A proper analysis of enormous data is required to get fruitful results. Classification, association rule mining, sequence analysis, and predictions are the significant outcomes of the data mining process. In the data mining process, classification is a widely used method. It is used to classify the data based on constraints and group the data in the dataset to predict future data labels. Classification technique can effectively handle the large dataset into a class label or groups so that samples with maximum similarity remain in the same set.

7.3 PROPOSED METHOD

To predict heart diseases, the Cleveland dataset was used. The dataset consists of age, gender, chest pain type, which can be non-anginal pain, typical angina, asymptomatic, atypical angina, resting blood pressure, resting electrocardiographic results, serum cholesterol in mg/dl results, ST-T wave abnormality or showing probable or definite left ventricular hypertrophy by Estes' criteria, fasting blood sugar, maximum heart rate achieved, resting heart rate; exercise-induced angina, ST depression induced by exercise relative to rest, the slope of the peak exercise ST-segment which can be flat, down or up sloping, number of major vessels (0–3) colored by fluoroscopy and diagnosis of heart disease (angiographic disease status) which showcases if the

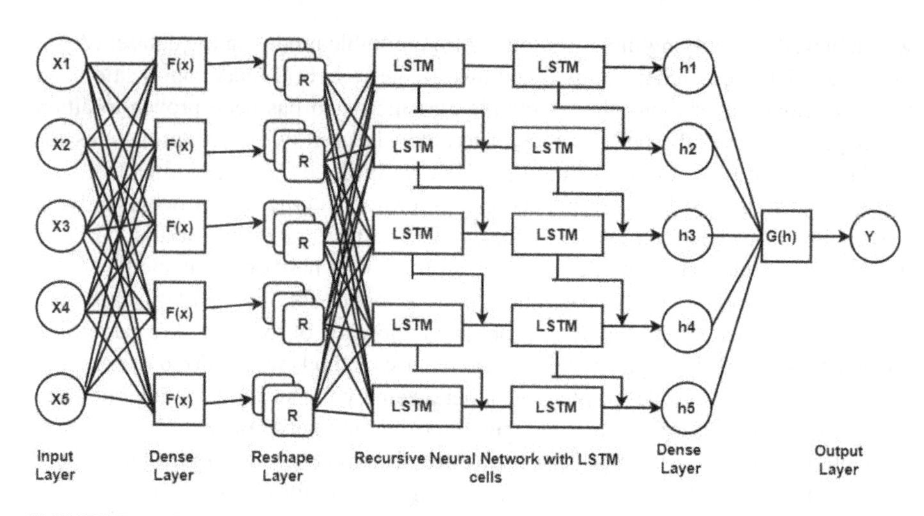

FIGURE 7.3 Proposed hybrid classification architecture.

diameter of the veins is narrowing or not. The model shown in Figure 7.3 was applied using the dataset, and heart diseases were predicted.

The proposed architecture works using the following steps:

- Input features are given to the system; these features are taken directly from the Cleveland dataset.
- Each of these features is given to a dense layer, wherein feature classification is done, and each of the features is reshaped.
- The reshape layer performs the task of grouping these features into different feature sets of similar values.
- These feature sets are stored in an LSTM network that performs a different combination of these features.
- All these combinations are given to an RNN, where the final classification process is performed using a dense layer and an output layer.
- The LSTMs and the RNN learn from each other during the entire process and minimize the classification error.

Each of the Cleveland dataset attributes were given a particular weight, and based on that weight, the final classification was performed. The results of this classification have been tabulated and compared with other advanced methods. The comparison is made in the next section of this chapter. It is observed that the algorithm under study performs optimally and can be used for real-time classification problems.

7.4 PERFORMANCE OF THE HYBRID CLASSIFICATION MODEL (RNN+LSTM) ON THE CLEVELAND DATASET

The RNN with LSTM hybrid algorithm has been compared with other algorithms like artificial neural networks, convolutional neural networks, and the like. Table 7.4 showcases this comparison.

TABLE 7.4
Comparison of Accuracies for Different Hybrid Classifiers

Classification model	Number of features used	Accuracy (%)
Stacking (Miškovic and Vladislav. 2014)	75	89
Multi Decision Trees (Kumar, Biswas and Walker 2020, 634)	62	91
NFE (Masabo et al. 2019)	68	93
GFS-LogitBoost-C (Abdeldjouad et al. 2020)	70	94
LRDA-GNN (Zhang and Verma 2005)	65	93
GABC (Muthuvel, Anto and Alexander 2019, 35372)	46	93
PSO-SVM (Kopiec and Martyna 2011)	59	94
Hybrid Transfer Learning (Kudva, K and Guruvare 2020, 625)	70	91
HCFC (Xiao et al. 2020, 2185)	34	94
NB-SVM (Ingole, Bhoir and Vidhate 2018, 14)]	14	95
GWO-CNN (Khan 2020, 34722)	18	94
RNN with LSTM (Proposed Model)	**13**	**96**

The RNN with LSTM classifier only utilizes 13 features from the Cleveland dataset. These features were selected from the given input set of more than 300 instances. Evaluation indicated that only the following parameters are necessary for the prediction:

1. Age of the patient (Age)
2. Gender of a patient (Gender)
3. Type of chest pain experienced by the patient (CP)
4. Resting blood pressure (Tresbps)
5. Serum cholesterol (Chol)
6. Fasting blood sugar (FBS)
7. Resting electrocardiographic results (Rest ECG)
8. Maximum heart rate of the patient (ThalACH)
9. Presence or absence of exercise-induced angina (ExANG)
10. ST depression induced by exercise relative to the rest (Old Peak)
11. Slope of peak exercises ST segment (Slope)
12. Number of significant vessels colored by fluoroscopy (CA)
13. Defect type of the patient (Thal)

They were the most relevant features because they produced maximum variations across different heart diseases. Moreover, due to the selection of these features using LSTM, RNN can classify the patients with heart disease and achieved an accuracy of more than 96% across training and testing values. It was also observed that the RNN accuracy can be perfected with a higher number of data instances because the

TABLE 7.5
Hybrid RNN and LSTM Specification

Simple RNN layer(100 units)
2 LSTM layer(100 units)
Activation-softmax
Optimizer – Adam
Epoch-100
Batch size-4

TABLE 7.6
Performance Metrics of the Proposed Hybrid Model for Heart Disease Prediction

Proposed hybrid algorithm	Accuracy	Precision	Recall	F-measure
RNN (Baviskar, Verma and Chatterjee 2020)	88	88	91	89
LSTM (Baviskar, Verma and Chatterjee 2020)	86	88	88	88
RNN + LSTM (Proposed Method)	96	94	97	95

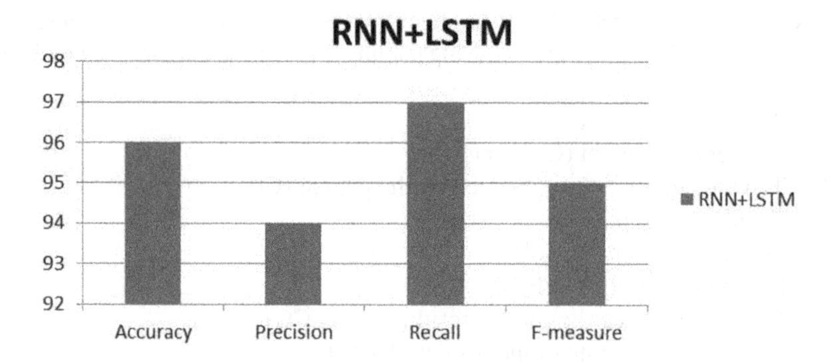

FIGURE 7.4 Hybrid algorithm with all performance metrics.

higher number of observations results in better neural network training. Therefore, it is recommended that LSTM with RNN be used as a network of choice to classify heart disease data sets. The detailed specification of the proposed hybrid model is shown in Table 7.5.

The following Table 7.6 shows various performance metrics of the proposed hybrid model on the Cleveland dataset

Graphical representation of the proposed algorithm with all performance metrics is shown in Figure 7.4.

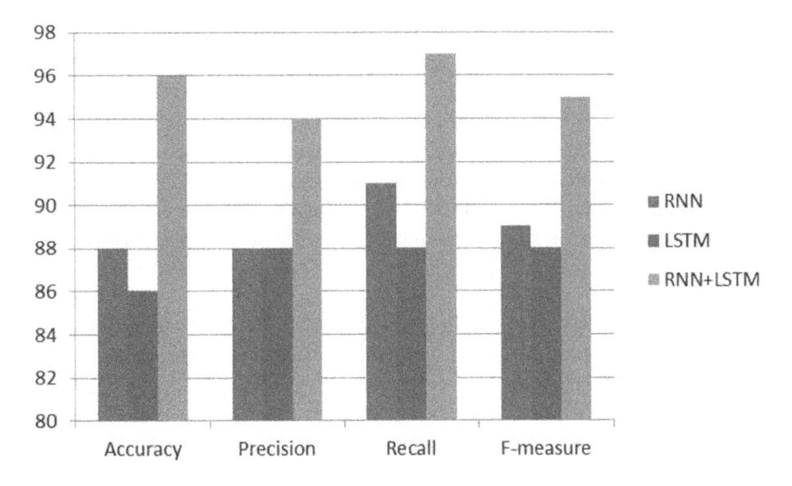

FIGURE 7.5 Hybrid algorithm comparison with DL algorithms.

Also, the graphical representation of the DL algorithms RNN and LSTM performance comparison is shown in Figure 7.5 with our proposed algorithm

As shown in Figure 7.4 and 7.5, a hybrid model of RNN and LSTM shows an accuracy of 96% on the Cleveland dataset.

7.5 CONCLUSION

Based on our review, it can be observed that CNN and RNN based methods are beneficial for the classification of heart diseases. The proposed model could efficiently perform the heart disease prediction on the Cleveland dataset. The feature selection technique selected only 13 features and the hybrid model (RNN+LSTM) was able to achieve an accuracy of 96%, which was compared and found to be better than other existing models. This classification accuracy can be further improved with deep-nets and 3D CNN models, which assist in effective feature extraction and classification.

REFERENCES

Abdeldjouad, Fatma, Brahami Menaouer, and Matta Nada. 2020. *A Hybrid Approach for Heart Disease Diagnosis and Prediction Using Machine Learning Techniques*. In: M. Jmaiel, M. Mokhtari, B. Abdulrazak H. Aloulou, and S. Kallel (eds), *The Impact of Digital Technologies on Public Health in Developed and Developing Countries*. ICOST 2020. Lecture Notes in Computer Science, vol 12157. Springer, Cham.

Agarwal, Manjari and Gaurav Kumar Ameta. 2019. "Implementation of an efficient hybrid classification model for heart disease prediction", *International Journal of Scientific & Technology Research*, vol. 8, no. 08, 292–297, August.

Asghar, Muhammad Zubair, Asmat Ullah, Shakeel Ahmad, and Aurangzeb Khan. 2020. "Opinion spam detection framework using hybrid classification scheme", *Soft Computing*, vol. 24, 3475–3498. doi.org/10.1007/s00500-019-04107-y.

Ashraf, Mohd., M. A. Rizvi and Himanshu Sharma 2019. "Improved heart disease prediction using deep neural network", *Asian Journal of Computer Science and Technology*, vol. 8, no. 2, 49–54 .

Baccouche, Asma, Begonya Garcia-Zapirain, Cristian Castillo Olea, and Adel Elmaghraby. 2020. "Ensemble deep learning models for heart disease classification: A case study from Mexico", *Information*, vol. 11, no. 4, 207. doi.org/10.3390/info11040207.

Bashir, Tariq, Imran Usman, Shahnawaz Khan, and Junaid Ur Rehman. 2017. "Intelligent reorganized discrete cosine transform for reduced reference image quality assessment", *Turkish Journal of Electrical Engineering & Computer Sciences*, vol. 25, no. 4, 2660–2673.

Baviskar, Vaishali, Madhushi Verma, and Pradeep Chatterjee. 2021. A Model for Heart Disease Prediction Using Feature Selection with Deep Learning. In: Deepak Garg, Kit Wong, Jagannathan Sarangapani, and Suneet Kumar Gupta (eds), *Advanced Computing. IACC 2020. Communications in Computer and Information Science*, vol. 1367. Springer, Singapore. doi.org/10.1007/978-981-16-0401-0_12.

Bouaziz, Fatiha, and Daoud Boutana. 2019. "Automated ECG heartbeat classification by combining a multilayer perceptron neural network with enhanced particle swarm optimization algorithm", *Research on Biomedical Engineering*, vol. 35, 143–155. doi.org/10.1007/s42600-019-00016-z.

Garg, S., K. Kaur, N. Kumar, G. Kaddoum, A. Y. Zomaya, and R. Ranjan. 2019. "A hybrid deep learning-based model for anomaly detection in cloud datacenter networks", *IEEE Transactions on Network and Service Management*, vol. 16, no. 3, 924–935, Sept. doi: 10.1109/TNSM.2019.2927886.

Ingole, Priyanka, Smita Bhoir, and A. V. Vidhate 2018. "Hybrid model for text classification," *2018 Second International Conference on Electronics, Communication and Aerospace Technology (ICECA). IEEE,* Coimbatore, India, pp. 7–15. doi: 10.1109/ICECA.2018.8474738.

Khan, Mohammad Ayoub. 2020. "An IoT framework for heart disease prediction based on MDCNN classifier", *IEEE Access*, vol. 8, 34717–34727. doi: 10.1109/ACCESS.2020.2974687.

Khatal, Sunil S. and Yogesh Kumar Sharma. 2020. "Analyzing the role of heart disease prediction system using IoT and machine learning", *International Journal of Advanced Science and Technology*, vol. 29 no. 9s, 2340–2346. sersc.org/journals/index.php/IJAST/article/view/14830.

Kopiec, Dawid, and Jerzy Martyna. 2011. A Hybrid Approach for ECG Classification Based on Particle Swarm Optimization and Support Vector Machine. In: Emilio Corchado, Marek Kurzyński, and Michał Woźniak (eds) *Hybrid Artificial Intelligent Systems*. HAIS 2011. *Lecture Notes in Computer Science*, vol. 6678. Springer, Berlin, Heidelberg. doi.org/10.1007/978-3-642-21219-2_42.

Kudva, Vidya, Keerthana Prasad, and Shyamala Guruvare. 2020. "Hybrid transfer learning for classification of uterine cervix images for cervical cancer screening", *Journal of Digital Imaging,* Jun., vol. 33, no.3, 619–631. doi: 10.1007/s10278-019-00269-1. PMID: 31848896; PMCID: PMC7256135.

Kumar, Jai, Brototi Biswas, and Sakshi Walker. 2020. "Multi-temporal LULC Classification using hybrid approach and monitoring built-up growth with Shannon's Entropy for a semi-arid region of Rajasthan, India", *Journal of the Geological Society of India*, vol. 95, 626–635. doi.org/10.1007/s12594-020-1489-x.

Miao, Kathleen H. and Julia H. Miao 2018. "Coronary heart disease diagnosis using deep neural networks", *International Journal of Advanced Computer Science and Applications* (IJACSA), vol. 9 no.10, dx.doi.org/10.14569/IJACSA.2018.091001.

Masabo, Emmanuel, Kyanda Kaawaase, Julianne Sansa-Otim, John Ngubiri, and Damien Hanyurwimfura. 2019. "Improvement of malware classification using hybrid feature engineering", *SN Computer Science*, vol.1, pp. 1–14. 10.1007/s42979-019-0017-9.

Miškovic, Vladislav. 2014. *Machine Learning of Hybrid Classification Models for Decision Support.* In: Sinteza 2014 – Impact of the Internet on Business Activities in Serbia and Worldwide, Belgrade, Singidunum University, Serbia, 2014, pp. 318–323. doi:10.15308/sinteza-2014-318-323.

Mohan, Senthilkumar, Chandrasegar Thirumalai, and Gautam Srivastava. 2019. "Effective heart disease prediction using hybrid machine learning techniques", *IEEE Access*, vol. 7, 81542–81554. doi: 10.1109/ACCESS.2019.2923707.

Muthuvel, K., S. Anto, and T. Jerry Alexander. 2019. "GABC based neuro-fuzzy classifier with hybrid features for ECG Beat classification", *Multimedia Tools and Applications*, vol. 78, 35351–35372. doi.org/10.1007/s11042-019-08132-9.

Nikookar, Elham. and Ebrahim Naderi 2018. "Hybrid ensemble framework for heart disease detection and prediction", *International Journal of Advanced Computer Science and Applications* (IJACSA), vol. 9, no. 5, dx.doi.org/10.14569/IJACSA.2018.090533.

Pillai, N. Sowri Raja, K. Kamurunnissa Bee, and J. Kiruthika. 2019. "Prediction of heart disease using RNN algorithm", *International Research Journal of Engineering and Technology* (IRJET) vol. 6, no. 3, 4452–4458, Mar 2019 www.irjet.net.

Pławiak, Paweł, and U. Rajendra Acharya. 2020. "Novel deep genetic ensemble of classifiers for arrhythmia detection using ECG signals", *Neural Computing and Applications*, vol. 32, 11137–11161. doi.org/10.1007/s00521-018-03980-2

Sharma, Sumit, Mahesh Parmar 2020. "Heart diseases prediction using deep learning neural network model", *International Journal of Innovative Technology and Exploring Engineering* (IJITEE), vol. 9, no. 3, 809–819, January 2020.

Singh, Navdeep and Sonika Jindal. 2018. "Heart disease prediction system using hybrid technique of data mining algorithms", *International Journal of Advance Research, Ideas and Innovation in Technology*, vol. 4, no. 2, 982–987.

Tarawneh, Monther, and Ossama Embarak. 2019. Hybrid Approach for Heart Disease Prediction Using Data Mining Techniques. In: Leonard Barolli, Fatos Xhafa, Zahoor Ali Khan, and Hamad Odhabi (eds) *Advances in Internet, Data and Web Technologies.* EIDWT 2019. *Lecture Notes on Data Engineering and Communications Technologies*, vol. 29. Springer, Cham. doi.org/10.1007/978-3-030-12839-5_41.

Torabi, Mehrnoosh, Sattar Hashemi, Mahmoud Reza Saybani, Shahaboddin Shamshirband, and Amir Mosavi. 2018. "A hybrid clustering and classification technique for forecasting short-term energy consumption, environmental progress & sustainable energy", *A Environmental Progress & Sustainable Energy*, vol. 38, 66–76. doi.org/10.1002/ep.

Udhaya Kumar, S., and H. Hannah Inbarani. 2015. Classification of ECG Cardiac Arrhythmias Using Bijective Soft Set. In: Hassanien, Aboul Ella, Ahmad Taher Azar, Vaclav Snasael, Janusz Kacprzyk, and Jemal H. Abawajy, Abawajy J. (eds) *Big Data in Complex Systems. Studies in Big Data*, vol. 9. Springer, Cham. doi.org/10.1007/978-3-319-11056-1_11.

Ul Haq, Amin, Jian Ping Li, Muhammad Hammad Memon, Shah Nazir, and Ruinan Sun 2018. "A hybrid intelligent system framework for the prediction of heart disease using machine learning algorithms", *Mobile Information Systems*, vol. 2018, Article ID 3860146, 21 pages, doi.org/10.1155/2018/3860146

Xiao, Jin, Yuhang Tian, Ling Xie, Xiaoyi Jiang, and Jing Huang. 2020. "A hybrid classification framework based on clustering", *IEEE Transactions on Industrial Informatics*, vol. 16, no. 4, 2177–2188, April 2020, doi: 10.1109/TII.2019.2933675.

Xiao Liu, Xiaoli Wang, Qiang Su, Mo Zhang, Yanhong Zhu, Qiugen Wang, Qian Wang. 2017. "A hybrid classification system for heart disease diagnosis based on the RFRSm",

Computational and Mathematical Methods in Medicine, vol. 2017, Article ID 8272091, 11 pages, doi.org/10.1155/2017/8272091.

Yang, Li, Haibin Wu, Xiaoqing Jin, Pinpin Zheng, Shiyun Hu, Xiaoling Xu, Wei Yu, and Jing Yan. 2020. "Study of cardiovascular disease prediction model based on random forest in eastern China", *Scientific Reports* vol. 10, 5245. doi.org/10.1038/s41598-020-62133-5.

Zhang, Ping, Kuldeep Kumar, and Brijesh Verma. 2005. A Hybrid Classifier for Mass Classification with Different Kinds of Features in Mammography. In: Lipo Wang and Yaochu Jin (eds) *Fuzzy Systems and Knowledge Discovery*. FSKD 2005. *Lecture Notes in Computer Science*, vol. 3614, pp. 316–319. Springer, Berlin, Heidelberg. doi.org/10.1007/11540007_38.

8 Intrusion Detection Using Hybrid Long Short-Term Memory with Binary Particle Swarm Optimization for Cloud Computing Systems

Hamza Turabieh and Noor Abu-el-rub

CONTENTS

8.1 INTRODUCTION

Nowadays, cloud computing systems help enterprises to operate several applications and store data over the internet. All provided services are remotely accessible over the internet, therefore, cloud computing systems help end users to overcome all issues related to installing, maintaining, and securing their data or applications in a convenient manner [Dey et al. 2019, Bridges et al. 2019]. Hence, it is critical to secure these cloud systems from abnormal traffic. Recently, many Cloud Service Providers (CSP) have investigated different methodologies to identify a robust IDS system to prevent illegitimate entry to access cloud systems. IDS operates by analyzing the

DOI: 10.1201/9781003138037-8

incoming traffic and classifying normal versus abnormal activities. Sometimes IDS blocks IP addresses that are related to unknown end users [Ghosh et al. 2019].

8.2 IDS METHODS

In general, there are two common methods for IDS: the host intrusion detection system (HIDS) and the network intrusion detection system (NIDS) [Vinayakumar et al. 2019]. HIDS controls and monitors all kinds of intrusions inside the local system, it keeps tracking and analyzing the traffic and all the information that comes from local machines to detect any abnormal behavior [Subba et al. 2021]. NIDS monitors all incoming traffic over the network to identify illegitimate activity. Both categories inform the system administrator about any possible attack that could happen or already have happened, in addition to performing a set of activities to prevent such attacks on the system [Ghosh et al. 2019].

IDS detects abnormal traffic by examining every packet and then takes an intelligent decision to keep cloud systems in a healthy status [Jyothsna et al. 2016]. IDS have two detection methods: anomaly and misuse (signature) based detection. The anomaly method tries to discover any abnormal traffic that deviates from the normal one. While misuse tries to detect abnormal based on previous known abnormal traffic patterns [Gamage et al. 2020]. In general, intrusion detection is a NP-Hard problem [Ravale et al. 2015; Elmasry et al. 2020], which can be solved using evolutionary computing and meta-heuristic methods.

Figure 8.1 explores the basic components of IDS inside the cloud computing system. IDS consists of four main functions which are: the packet processing function, the feature reduction function, the ML classifier function, and feature selection. The packet processing function handles the incoming packets from the network and converts these packets into a standard format as row data. The feature reduction function is used to improve the collected data (i.e., row data) by removing redundant and missing data. The reduction data is used as input data to a ML classifier, where the output of the ML classifier is used to update the IDS database and notify the system if an abnormal packet is received. The feature selection function can be employed to enhance the overall performance of the classifier by having a lower dimensionality of the data.

Any IDS should have three fundamental features to secure the cloud computing system which are: data confidentiality, data integrity, and data availability [Patel et al. 2020]. Data confidentiality refers to sensitive data that cannot be interrupted by untrusted users. Data integrity means data tampering should not be allowed during the transmission process. Data availability ensures that the network resources and data can always be accessed by end users.

There are several existing IDS that try to build an intelligent classification system based on a set of historical data. Since network traffic data is considered high dimensional data, several researchers employ feature selection (FS) methods to enhance the data quality and reduce the data dimensionality [Alazzam et al. 2020]. FS methods enhance the overall performance of ML [Prasad et al. 2020]. For example, Almomani [2020] used four different types of FS methods namely: genetic algorithm (GA),

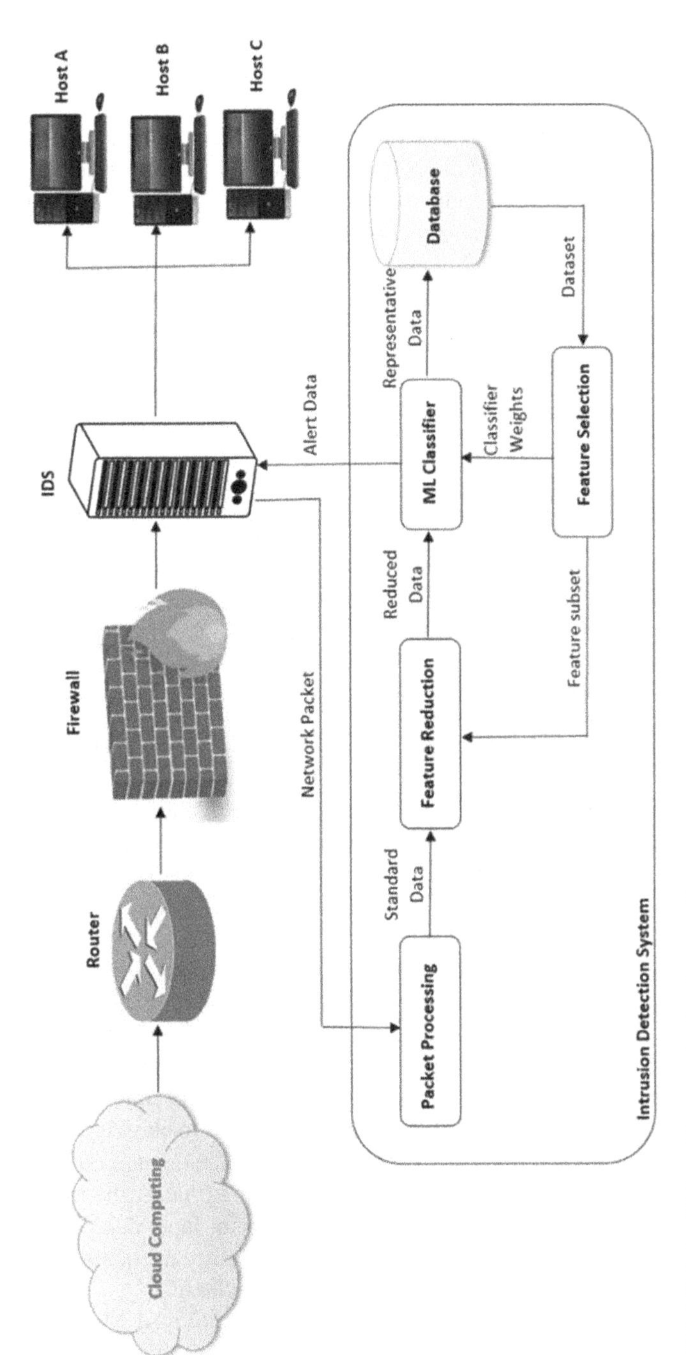

FIGURE 8.1 IDS architecture in cloud computing.

particle swarm optimization (PSO), firefly optimization (FFA), and grey wolf optimizer (GWO). Almomani in his work showed that FS methods can enhance the performance of ML classifiers such as support vector machine (SVM) and decision tree (J48). Another example of employing FS methods with IDS was proposed by Thakkar and Lohiya [2020], where the authors applied seven ML classifiers (i.e., neural networks (NN), decision tree (DT), logistic regression (LR), support vector machine (SVM), k-nearest neighbors (kNN), random forest (RF), and naïve Bayes (NB)). Moreover, the authors applied a two-wrapper feature selection namely: Chi-squared, information gain (IG), and recursive feature elimination (RFE). The obtained results show an excellent performance of FS methods with ML classifiers. Zhu et al. [2017] introduced a multi-objective method for FS for building a robust IDS inside cloud computing systems.

In this chapter, the main research contributions are listed below.

1. Investigate the performance of wrapper feature selection to reduce the dimensionality of the data.
2. Investigate the performance of LSTM as a deep learning method to classify incoming packets.

The rest of this chapter is organized as follows: Section 2 summarizes the related work of ML and FS in the area of IDS in cloud computing, Section 3 shows the proposed method, Section 4 introduces the dataset used in this chapter (that is, UNSW-NB15), Section 5 presents the obtained results and provides a deep discussion, and finally, Section 6 shows the findings of this chapter and suggests future work.

8.3 RELATED WORKS

With the increase of using cloud computing, more attention is given to network security to ensure application and data security. The most important aspect is to quickly detect attacks once they have occurred and take the correct actions to protect the cloud systems. In general, there are five types of network attacks: DOS, U2R, NORMAL, R2L, and PROBE [De la Hoz 2014]. The collected data is generated from a huge number of packets. As a result, the number of samples (that is, records) is large, which make it difficult to process this data. To overcome this problem, feature selection has been used widely to enhance the IDS.

Machine learning has several methods that can be used in IDS such as: SVM, DT, ANN, CNN, and so forth. All of these methods use a learning algorithm to learn from previous data that has been classified. Basically, ML tries to classify training data by minimizing the classification errors between actual and estimated output. Later, the trained model will be evaluated based on collected data (that is, the testing dataset). The performance of ML depends on three things: type of input data, quality of input data, and learning algorithm. Feature selection will enhance the quality of input data, which will reflect on the overall performance of a ML classifier.

There are many published research papers that employed ML and FS in the detection of abnormal packets. For example, Aneetha et al. [2012] proposed an intelligent analysis approach to detect intrusions. The proposed method used clustering methods with rules based on classifying incoming packets. The reported results show that a rule based cluttering method can enhance the performance of IDS. Sun et al. [2017] proposed a cloud based anti-malware system called Cloud Eyes. The proposed system can provide a high-quality, cost-effective service for end users by protecting their data and applications. Moreover, Cloud Eyes can secure cloud servers and cloud clients. Shen et al. [2018] implemented a malware detection infrastructure for IDS for cloud computing systems. The main objective was to detect all incoming abnormal packets. Sivatha et al. [2012] introduced a lightweight intrusion detection system that combines three algorithms that are: the genetic algorithm, neural networks, and a decision tree. The genetic algorithm (GA) is employed as a wrapper feature selection, while neural networks, and decision trees are combined to generate a hybrid classifier called Neurotree. The proposed algorithm evaluates a public dataset called NSL-KDD. The performance of GA as a feature selection, was great, with it selecting 14 features out of 41. The Neurotree classifier shows an excellent performance based on accuracy value (it is 98.38%).

Shahri et al. [2016] combines the genetic algorithm with SVM to detect intrusions. The authors employed the GA as a feature selection, with SVM as a ML classifier. In this work, the authors simulate their proposed method over the KDD'99 dataset. The dataset consists of 41 features. The feature selection algorithm reduces the number of features to 10. The performance of SVM was able to detect abnormal packets with accuracy equals 97.3%.

Selvakumar et al. [2019] proposed a swarm optimization algorithm, called the Firefly algorithm as a feature selection method. The authors simulate their proposed method over the KDDCUP 99 dataset. The obtained results show that swarm algorithms work well as feature selection algorithms. AlYaseen [2019] applied the same algorithm, the firefly algorithm, as a feature selection with the SVM classifier. The author reports that the performance of this proposed approach shows an excellent performance with accuracy of 78.89%. Yang et al. [2009] investigated the importance of feature selection algorithms for IDS. In this work, the authors proposed a lightweight IDS based on modified haphazard mutation hill climbing. Yinhui et al. [2012] employed a feature selection algorithm called, the gradually feature removal method (GFR). The authors tested their algorithm over the KDD Cup dataset. The GFR selects 19 features out of 41. The authors applied SVM to examine the selected features, where the performance of SVM is 98.62%.

This chapter examines a wrapper feature selection based on BPSO and LSTM as a classifier. It aims to employ the proposed method over the UNSW-NB15 dataset.

8.4 PROPOSED APPROACH

8.4.1 Binary PSO

In 1995, Kennedy and Eberhart [1995] proposed a swarm intelligence algorithm called particle swarm optimization (PSO). Simply put, PSO simulates the behavior of organisms while flying. PSO is a population-based solution, where all solutions are moving in the search space. Each solution x (that is, particle) has two values: position x_{id} and velocity v_{id}. Both variables are calculated based on the location of the best

```
Given:
    -Sn: swarm size.
    -t: number of iterations.
    -v: initial velocity.
    -x: initial position.
    -c1: degree of influence of p_id.
    -c2: degree of influence of p_gd.
initialize particle()
While (current_iteration ≤ t)
    Evaluate each particle's position according to the fitness function.
     Find the best solution of each particle so far.
     update the global best solution.
     update the velocity of each particle based on Eq. 1.
     update the position of each particle based on Eq. 2.
end While
Output the global best solution
```

FIGURE 8.2 The pseudo-code for particle swarm optimization.

solution p_i in the search space, and the location of the best solution in the neighborhood p_g. Where i presents the solution number in the population

($1=1, 2,\ldots,S_n$), n is the size of population, and t refers to the number of iterations. Eq. (1) and Eq. (2) presents the updating approach for both position x_{id} and velocity v_{id}, respectively. Where the numbers r_1 and r_2 represent two random numbers in the range [0,1], w represents a positive inertia weight, c_1 refers to the degree of influence of p_{id}, while c_1 refers to the degree of influence of p_{gd}. To make sure that all solutions are in feasible locations in the search space, the velocity is controlled within a range $[v_{min}, v_{max}]$. Figure 8.2 explores the pseudo-code for particle swarm optimization.

$$v_{id}(t+1) = wv_{id}(t) + c_1 r_1 [p_{id}(t) - x_{id}(t)] + c_2 r_2 [p_{gd}(t) - x_{id}(t)]. \quad (1)$$

$$x_{id}(t+1) = x_{id}(t) + v_{id}(t+1). \quad (2)$$

In this chapter, we adopt the concept of Binary PSO (BPSO), where the solutions can either be 0 or 1 (that is, discrete). To switch continuous PSO to BPSO, a transfer function (TF) is needed. Here, a sigmoid transfer function is used, based on Eq. (3).

$$S\left(v_{id}(t+1)\right) = \frac{1}{1+e^{-v_{id}(t)}} \quad (3)$$

where V_{id} represents the velocity value of the d_{th} dimension in the ith vector, and t for the current iteration. The updating approach for the current solution is shown in Eq. (3). Where the variable x_{id} represents the element in the dth dimension in the ith position in the next iteration, randis a random function-generator for a random number between [0,1].

$$x_{id}(t+1) = \begin{cases} 1 & \text{if } rand(0.0,1.0) < S(v_{id}(t+1)) \\ 0 & otherwise \end{cases} \tag{4}$$

8.4.2 Long Short-Term Memory (LSTM) Networks

Here, a deep learning method was employed (that is, CNN-LSTM) to detect intrusions. Figure 8.3 explores the main structure of CNN-LSTM. Basically, LSTM uses internal memory to memorize the temporal sequence of the input feature vectors.

LSTM tries to map the input i (that is, features) with output o (that is, abnormal/normal packet), while forgetting the f gate to memorize the store features. The hidden state h cell state c are used for memorizing. All calculations of LSTM are shown in Eqs. (5, 6, and 7).

$$\begin{pmatrix} i \\ f \\ o \\ g \end{pmatrix} = \begin{pmatrix} sigmoid \\ sigmoid \\ sigmoid \\ tanh \end{pmatrix} w^t \begin{pmatrix} h_t^{l-1} \\ h_{t-1}^{l-1} \end{pmatrix} + \begin{pmatrix} b_i \\ b_f \\ b_o \\ b_g \end{pmatrix} \tag{5}$$

$$c_t = f_t \circ c_{t-1} + i_t \circ g \tag{6}$$

$$h_t = o_t \circ \sigma(c_t) \tag{7}$$

The calculation of a fully connected layer and softmax process are shown in Eq.(8), and Eq.(9), respectively. In this work, the softmax was employed to classify the input user's role. While the output of the fully connected layer is presented by the softmax layer in the range [0,1]. N_c refers to the number of rules, and L represents the activity class probability.

$$d_i^l = \sum_i \sigma\left(W_{ji}^{l-1}\left(h_i^{l-1}\right) + b_i^{l-1}\right) \tag{8}$$

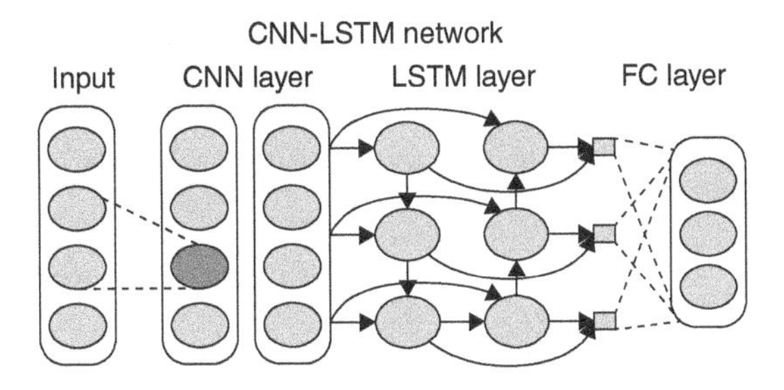

FIGURE 8.3 The main structure of CNN-LSTM.

$$P(c \mid d) = \arg\max_{c \in C} \frac{\exp\left(d^{L-1} w^{L}\right)}{\sum_{k=1}^{N_c} \left(d^{L-1} w_k\right)} \tag{9}$$

8.5 DATASET

In this chapter, the proposed hybrid approach over a public intrusion dataset called UNSW-NB1, was evaluated. The dataset is generated using a tool called, IXIA PerfectStorm by Moustafa et al. [2015]. The dataset has 9 different types of attack. The dataset has 49 features. In this work, only 44 features are used.

Table 8.1 explores 44 features of the dataset. The UNSW-NB dataset is an imbalanced dataset. In this work the adaptive synthetic sampling method (ADASYN) was employed to solve the class imbalance issue [Haibo et al. 2008]. Table 8.2 explores the original and balanced dataset. In this work this dataset was used as a binary classification problem to determine normal or abnormal attacks

TABLE 8.1
Features of UNSW_NB15 Dataset

Feature number	Feature name	Type	Feature number	Feature number	Type
1	id	Nominal	23	dtcpb	Integer
2	dur	Float	24	dwin	Integer
3	proto	Nominal	25	tcprtt	Float
4	service	Nominal	26	synack	Float
5	state	Nominal	27	ackdat	Float
6	spkts	Integer	28	smean	Integer
7	dpkts	Integer	29	dmean	Integer
8	sbytes	Integer	30	trans_depth	Integer
9	dbytes	Integer	31	response_body_len	Integer
10	rate	Integer	32	ct_srv_src	Integer
11	sttl	Integer	33	ct_state_ttl	Integer
12	dttl	Integer	34	ct_dst__ltm	Integer
13	sload	Float	35	ct_src_dport_ltm	Integer
14	dload	Float	36	ct_src_sport_ltm	Integer
15	sloss	Integer	37	cr_dst_src_ltm	Integer
16	dloss	Integer	38	is_ftp_logon	Binary
17	sinpkt	Integer	39	ct_dtp_ltm	Integer
18	dinpkt	Integer	40	ct_src_ltn	Integer
19	sjit	Float	41	ct_srv_dst	Integer
20	djit	Float	42	ct_sm_ips_ports	Integer
21	swin	Integer	43	is_sm_ips_ports	Binary
22	stcpb	Integer	44	attack_cat	Nominal

TABLE 8.2
Original AND Balanced UNSW NB15 Dataset

Dataset	Number of Normal	Number of Attacks	Total
Original Training	56000	119341	175341
Original Testing	37000	45332	82332
Balanced Training	119341	119341	238682
Balanced Testing	45332	45332	90664

8.6 RESULTS AND ANALYSIS

This section reports the validation of the proposed hybrid method (that is, BPSO with LSTM) to detect intrusions in cloud computing systems. All experiments are undertaken, based on a cross validation method with kfold = 10. We implemented the proposed approach using MATLAB 2019b.

We used six criteria to evaluate the proposed method: accuracy (See Eq. (10), specificity (See Eq. (11), precision (See Eq. (13), recall (See Eq. (14), and the F-Measure (See Eq. (15)).

$$Accuracy = \frac{TP + TN}{TP + FP + FN + TN} \tag{10}$$

$$Specificity = \frac{TN}{TN + FP} \tag{11}$$

$$Sensitivity = \frac{TP}{TP + FN} \tag{12}$$

$$Precision = \frac{TP}{TP + FP} \tag{13}$$

$$Recall = \frac{TP}{TP + FN} \tag{14}$$

$$F - Measure = \frac{2 \times (Recall \times Precision)}{Recall + Precision} \tag{15}$$

In this work, the proposed hybrid approach was applied, with three settings: balanced dataset with FS (that is, BPSO), balanced dataset without FS, and the original dataset without FS. Table 8.3 explores the obtained results for three types of experiments. The performance of feature selection improves the overall performance of LSTM compared to other experiments without feature selection. For example, the obtained

TABLE 8.3
Obtained Results for the Proposed Method

	Balanced Dataset with FS		Balanced Dataset		Original Dataset	
	Training	Testing	Training	Testing	Training	Testing
Accuracy	0.97	0.92	0.95	0.84	0.86	0.86
Sensitivity	0.95	0.99	0.94	0.84	0.96	0.88
Specificity	1.00	0.84	0.99	0.85	0.71	0.82
Percision	1.00	0.88	1.00	0.84	0.82	0.85
Recall	0.95	0.99	0.94	0.84	0.96	0.88
F-measure	0.98	0.91	0.97	0.86	0.89	0.87

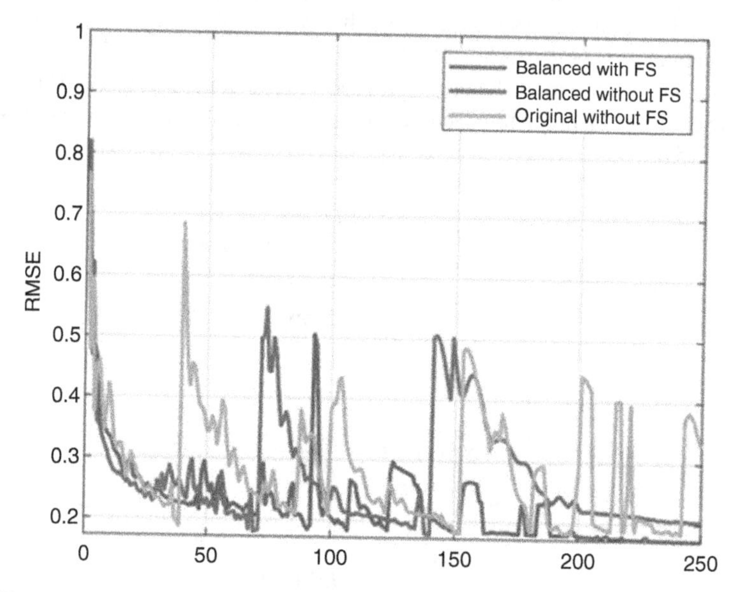

FIGURE 8.4 LSTM convergence for a training dataset based on RMSE.

results for the testing dataset show a good improvement (that is, 6%) for the proposed method over the balanced dataset. Figure 8.4 explores the performance of LSTM in the training process. The classification error (that is, RMSE) has a smooth convergence for balanced data with feature selection (that is, blue line).

From the obtained results, we believe that the proposed method can enhance the overall performance of IDS inside a cloud computing system.

8.7 CONCLUSION

In this chapter, we highlight the importance of feature selection methods for enhancing the performance of ML classifiers for cloud computing systems. Many companies start using this technology to save time, cost, and enhance security for their applications and data. Intrusion detection systems (IDS) play a vital role in protecting cloud platforms. In this work, a hybrid method was employed, between BPSO as a wrapper feature selection method and LSTM as a ML classifier. The proposed method was employed to detect intrusion. The proposed approach was examined over a public dataset, called UNSW-NB15. The results obtained show a good performance of detecting incoming intrusions with an accuracy of 92%.

REFERENCES

Alazzam, Hadeel, Ahmad Sharieh, and Khair Eddin Sabri. "A feature selection algorithm for intrusion detection system based on Pigeon Inspired Optimizer." *Expert Systems with Applications* 148 (2020): 113249. doi.org/10.1016/j.eswa.2020.113249. www.sciencedirect.com/science/article/pii/S0957417420300749.

Almomani, Omar. "A feature selection model for network intrusion detection system based on PSO, GWO, FFA and GA algorithms." *Symmetry* 12, no. 6 (2020): 1046.

Al-Yaseen, Wathiq Laftah. "Improving intrusion detection system by developing feature selection model based on Firefly algorithm and support vector machine." *IAENG International Journal of Computer Science* 46, no. 4 (2019).

Aneetha, A. S., T. S. Indhu, and S. Bose. "Hybrid network intrusion detection system using expert rule based approach." In *Proceedings of the Second International Conference on Computational Science, Engineering and Information Technology*. Association for Computing Machinery, New York. NY, (2012): 47–51.

Aslahi-Shahri, B. M., Rasoul Rahmani, M. Chizari, A. Maralani, M. Eslami, Mohammad Javad Golkar, and A. Ebrahimi. "A hybrid method consisting of GA and SVM for intrusion detection system." *Neural computing and applications* 27, no. 6 (2016): 1669–1676.

Bridges, Robert A., Tarrah R. Glass-Vanderlan, Michael D. Iannacone, Maria S. Vincent, and Qian Chen. "A survey of intrusion detection systems leveraging host data." *ACM Computing Surveys* (CSUR) 52, no. 6 (2019): 1–35

De la Hoz, Emiro, Eduardo De La Hoz, Andrés Ortiz, Julio Ortega, and Antonio Martınez-Álvarez. "Feature selection by multi-objective optimisation: Application to network anomaly detection by hierarchical self-organising maps." *Knowledge-Based Systems* 71 (2014): 322–338.

Dey, Saurabh, Qiang Ye, and Srinivas Sampalli. "A machine learning based intrusion detection scheme for data fusion in mobile clouds involving heterogeneous client networks." *Information Fusion* 49 (2019): 205–215. doi.org/10.1016/j.inf fus.2019.01.002. www.sciencedirect.com/science/article/pii/S1566253518306110.

Elmasry, Wisam, Akhan Akbulut, and Abdul Halim Zaim. "Evolving deep learning architectures for network intrusion detection using a double PSO metaheuristic." *Computer Networks* 168 (2020): 107042.

Gamage, Sunanda, and Jagath Samarabandu. "Deep learning methods in network intrusion detection: A survey and an objective comparison." *Journal of Network and Computer Applications* 169 (2020): 102767.

Ghosh, Partha, Arnab Karmakar, Joy Sharma, and Santanu Phadikar. "CSPSO based Intrusion Detection System in Cloud Environment." In *Emerging Technologies in Data Mining and Information Security*, edited by Ajith Abraham, Paramartha Dutta, Jyotsna Kumar Mandal, Abhishek Bhattacharya, and Soumi Dutta, (2019): 261–269. Singapore: Springer Singapore.

Haibo He, Yang Bai, E. A. Garcia, and Shutao Li. "ADASYN: Adaptive synthetic sampling approach for imbalanced learning." In *2008 IEEE International Joint Conference on*

Neural Networks (IEEE World Congress on Computational Intelligence), 1322–1328. 2008. doi.org/10.1109/ IJCNN.2008.4633969.

Jyothsna, V., and V. V. Rama Prasad. "FCAAIS: Anomaly based network intrusion detection through feature correlation analysis and association impact scale." *Special Issue on ICT Convergence in the Internet of Things* (IoT), ICT Express 2, no. 3 (2016): 103–116. doi.org/10.1016/j.icte.2016.08.003.

Kennedy, James and Russell Eberhart. "Particle swarm optimization." In *Neural Networks, 1995. Proceedings., IEEE International Conference on,* vol. 4 , 1942–1948 vol.4. November 1995. doi.org/10.1109/ICNN.1995.488968.

Li, Yang, Jun-Li Wang, Zhi-Hong Tian, Tian-Bo Lu, and Chen Young. "Building light-weight intrusion detection system using wrapper-based feature selection mechanisms." *Computers & Security* 28, no. 6 (2009): 466–475.

Li, Yinhui, Jingbo Xia, Silan Zhang, Jiakai Yan, Xiaochuan Ai, and Kuobin Dai. "An efficient intrusion detection system based on support vector machines and gradually feature removal method." *Expert Systems with Applications* 39, no. 1 (2012): 424–430.

MATLAB and Statistics Toolbox Release 2012b, The MathWorks, Inc., Natick, Massachusetts, United States.

Moustafa, Nour and Jill Slay. "UNSW-NB15: a comprehensive data set for network intrusion detection systems (UNSW-NB15 network data set)." In *2015 Military Communications and Information Systems Conference* (MilCIS), (2015): 1–6. doi.org/10.1109/MilCIS.2015.7348942.

Patel, Reema, Amit Thakkar, and Amit Ganatra. "A survey and comparative analysis of data mining techniques for network intrusion detection systems." *International Journal of Soft Computing and Engineering* (IJSCE) 2, no. 1 (2012): 265–260.

Prasad, Mahendra, Sachin Tripathi, and Keshav Dahal. "An efficient feature selection based Bayesian and Rough set approach for intrusion detection." *Applied Soft Computing* 87 (2020): 105980.

Ravale, Ujwala, Nilesh Marathe, and Puja Padiya. "Feature selection based hybrid anomaly intrusion detection system using K means and RBF kernel function." *International Conference on Advanced Computing Technologies and Applications (ICACTA), Procedia Computer Science* 45 (2015): 428–435. doi.org/10.1016/j.procs.2015. 03.174. www. sciencedirect.com/science/article/pii/ S1877050915004172.

Selvakumar, B, and Karuppiah Muneeswaran. "Firefly algorithm based feature selection for network intrusion detection." *Computers & Security* 81 (2019): 148–155.

Shen, Shigen, Longjun Huang, Haiping Zhou, Shui Yu, En Fan, and Qiying Cao. "Multistage signaling game-based optimal detection strategies for suppressing malware diffusion in fog-cloud-based IoT networks." *IEEE Internet of Things Journal* 5, no. 2 (2018): 1043–1054.

Sindhu, Sivatha., Siva S., S. Geetha, and A. Kannan. "Decision tree based light weight intrusion detection using a wrapper approach." *Expert Systems with Applications* 39, no. 1 (2012): 129–141. doi.org/10.1016/j.eswa.2011.06.013. www.sciencedirect.com/science/article/pii/S0957417411009080.

Subba, Basant, and Prakriti Gupta. "A tfidfvectorizer and singular value decomposition based host intrusion detection system framework for detecting anomalous system processes." *Computers Security* 100 (2021): 102084.

Sun, Hao, Xiaofeng Wang, Rajkumar Buyya, and Jinshu Su. "CloudEyes: Cloudbased malware detection with reversible sketch for resource-constrained internet of things (IoT) devices." *Software: Practice and Experience* 47, no. 3 (2017): 421–441.

Thakkar, Ankit, and Ritika Lohiya. "Attack classification using feature selection techniques: a comparative study." *Journal of Ambient Intelligence and Humanized Computing* 12, no. 1, (2020): 1–18.

Vinayakumar, Ravi, Mamoun Alazab, K. P. Soman, Prabaharan Poornachandran, Ameer Al-Nemrat, and Sitalakshmi Venkatraman. "Deep learning approach for intelligent intrusion detection system." *IEEE Access* 7 (2019): 41525–41550. doi.org/10.1109/ACCESS.2019.2895334.

Zhu, Yingying, Junwei Liang, Jianyong Chen, and Zhong Ming. "An improved NSGA-III algorithm for feature selection used in intrusion detection." *Knowledge Based Systems* 116 (2017): 74–85.

9 A Novel Live Streaming Platform Using Cloud Front Technology

Proof of Concept for Real Time Concerts

Suja Panicker, Amit Nene, Ashish Hardas, Shraddha Kamble, and Kaustubh Bhujbal

CONTENTS

DOI: 10.1201/9781003138037-9

9.1 INTRODUCTION

In earlier times when the pandemic (Covid19) was not an issue we could casually go to theatres, attend colleges, visit places, restaurants and attend live concerts. However, everything dramatically during just six months. People were forced to stay at home, and they found alternatives for their normal activities. An example of this it that instead of going to college, students started attending lectures online. A similar scenario was enacted in other day to day life events. Eventually, live concerts were held online on platforms like Zoom, Google Meet, and the like. But issues like impaired audio and video quality were a difficulty. To conduct smooth online concerts and to improve audio and video quality the idea developed of creating a special video conferencing application. By considering the various alternatives, the idea of the OpenVidu platform emerged, which uses the concept of WebRTC for real time communication. Some innovations were made, and new features added, such as: audio video recording, chat functionality, and advance camera functionality. Thus, integrating other services like AWS and OBS a platform was developed that has features that many major applications currently lack.

9.2 MOTIVATION

Due to Covid 19 there has been an overriding need to conduct meetings, sessions, and events in online mode. Various applications exist that conduct video conferences using WebRTC technology which aids in the smooth conduction of online live streamed conferences. Most of these applications currently on the market are suitable for simple voice communication but don't support polyphonic. Polyphony means having a musical texture consisting of two or more voices simultaneously. This provided a motive to overcome the polyphony problem and to remove noise while conducting a live concert, hence the advent of this novel application.

This work is organized thus: Section 3 presents a literature review, Section 4 covers proposed work, Section 5 elaborates on technology used and experimentation, Section 6 presents applications, Section 7 presents novelty, Section 8 highlights research contribution, Section 9 covers future work and finally, there is the conclusion.

9.3 LITERATURE REVIEW

Videoconferencing is a technology that makes use of a communication medium allowing multiple connected users at geographically dispersed locations to share audio and video content in real time (Samarraie et al., 2019; Krutka and Carano, 2016). The use of videoconferencing has seen an ever-increasing growth worldwide, and this is primarily attributed to: easy availability of higher bandwidths, networks, and of high computing speed. Hence, videoconferencing has emerged as the provider of solutions for various organizations, schools, universities, and the like in the majority of developed/developing countries worldwide (Samarraie et al., 2019).

Although videoconferencing has been in existence since the 1960s, the high cost factor associated with it has made it inaccessible to a number of organizations worldwide (Sondak et al., 1995). There have been several interesting case studies on video conferencing in the field of education, in the pre-pandemic days (Al-Samarraie et al., 2019, Khalid and Hossan, 2016, Hampel et al., 2005); also in healthcare (Drude, 2020, Humer et al., 2020, Marhefk et al., 2020). However, the research problem that has been addressed in the current work is quite novel and will be significant for artists who are conducting concerts through real time video-conferencing.

An ontology of the field of cloud computing is presented vividly in (Youseff et al., 2008). Its authors claim it to be a novel contribution in the field that establishes a clear ontology of the cloud which will be significant for fellow researchers. In this work, cloud computing is classified into five main layers: hardware, software infrastructure, kernel, software environments, and applications.

The hardware layer underpins the other layers and is the physical element of the system. The application layer is topmost and serves as an interface between the various browsers and the users. It was noted that despite desktop as a service (DaaS) systems to lower latency during data transfer, data leakage still cannot be ruled out. Currently security mechanisms include the public key infrastructure and the X.509 SSL certificate for authentication/authorization in the cloud. Another important issue is that of data ownership. Current work addresses the tradeoffs and challenges at each of these layers. It was also observed that owing to lack of standards, ownership and data privacy policies are usually approached in different ways by different cloud providers. It is also noted that in contrast to some existing work (Crandell, 2008; Cloud Services Continuum, 2008; The Cloud Services Stack and Infrastructure, 2008; Sabahi et al., 2012) which provide a general classification, current work is more systematic, scientific and also addresses the research challenges (Youseff et al., 2008)

There have been several instances of research in this field. A significant project has been (Lombardia et al., 2010) which demonstrates virtualization to increase cloud security through the integration of protection mechanisms with virtual machines that are also components of cloud infrastructure. Also, an innovative architecture to provide increased security is proposed, deployed on multiple cloud solutions thereby monitoring integrity, while also being completely transparent for cloud users. A prototype is successfully implemented on the following solutions (open source): OpenECP, and Eucalyptus. Two important research objectives were: effectiveness and performance. Experimental results indicated attack resilience and it was observed that there is a small overhead that gets introduced in the process but can be overlooked in comparison to the numerous other benefits.

In (Edan et al., 2017) the WebSocket protocol was utilized in the communication occurring between two browsers. Also, WebRTC was used for video conferencing to provide bidirectional communication and experimentation took place on different network types. The performance of CPU, quality of experience (QoE) and bandwidth used were noted and used for benchmarking. Future work includes creating a signalling mechanism that moves towards unlimited peers and on application to further network topologies. Future work also covers the comparison of WebRTC to other popular protocols such as IAX2, SIP, and the like.

WebRTC has contributed a new functionality to existing web browsers thereby allowing audio/video calls between different browsers without any need for installation of video telephony. The congestion control in WebRTC is provided by the

Google congestion control technique. However, the latter has limited performance owing to the use of the constant incoming rate decrease factor. A dynamic model is presented for estimating the receiving bandwidth during sessions of very high usage. After sufficient experimentation it was noted that the use of a specific testbed helped achieve 33% increased incoming rate using the proposed model with 16% reduced time for the round-trip (Atwah et al., 2015)

In an attempt to gain insights on enhancing the QoE during videoconferencing using WebRTC, a pilot study was conducted in (Moor et al., 2017) with 22 subjects. Under differing technical conditions two-party audiovisual communications occurred during the study. Researchers collected self-reports, physiological data, and other statistics, and these were explored with respect to usefulness and compatibility for the efficient assessment of QoE. Preliminary results indicated that the differing quality is evident from self-reports. Future work includes considering a large sample size, benchmarking with physiological data, and in-depth investigation of other underlying factors.

A P2P system, based on WebRTC, for providing scalability in real time video conferencing has been experimented upon in (Apu et al., 2017), with results conclusively stating that there is the capability in WebRTC for scalable conferencing in browsers. An innovative multi feature video conferencing solution has been experimented upon in (Grozev et al., 2018). The concept of geo-location was used, and distributed mode was experimented with wherein one conference was split and distributed to multiple servers.

Table 9.1 presents highlights of latest research in this field.

As observed from Table 9.1, there are various technological constraints in currently existing applications, hence the usefulness of our proposed work which addresses these shortcomings. It is noted that key considerations in choosing a suitable online platform include: accessibility, availability, cost, user friendliness, and security (Connolly et al., 2020).

9.4 PROPOSED WORK

The proposed system focuses on providing a platform for performers to conduct online concert/programs with good audio and video quality.

In the proposed system (shown in Figure 9.1) the major focus is an improved video/audio quality to be provided in a teleconferencing application where a performer can showcase his/her talent on an online platform.

In the architecture (Figure 9.1) the performer joins the sessions through any cross platform. The technology used for establishing a teleconferencing session is OpenVidu. The OpenVidu server helps establish a good quality teleconferencing session. The session established through OpenVidu is further streamed into the OBS studio which is a free and open-source cross platform streaming and recording program built with Qt (a toolkit for developing graphical user interfaces). Once the stream is generated successfully via OBS studio we use various AWS services for establishing the live session using cloud front technology for access by the worldwide user.

TABLE. 9.1
Highlights of Literature Survey

Ref. No.	Authors	Publication year	Basic concepts used.	Advantages	Research gap
Xu et al., 2012	Hongfeng Xu, Zhen Chen,Junwei Cao	2012	Various streaming techniques are used - HTTP live streaming, RTSP, Adobe Flash etc.	1. Content Caching 2. Location-independence	1. Availability 2. Location-dependence
Sondak et al., 1995	Jiushuang Wang; Weizhang Xu; Jian Wang	2016	The incorporation of multimedia on mobile phones adds tremendous value to user communications owing to smarter networks, faster internet and latest streaming technologies.	1. Streaming video can increase your productivity 2. Video streaming allows flexibility	1. Videos aren't saved 2. Poor battery life
Smiti et al., 2018	Puja Smiti, Swapnita Srivastava Nitin Rakesh	2018	Various compression techniques are reviewed.	Enhancing the video transmission standard before refetching data for the mobile users.	Upgrading the video quality.
Stefan et al., 2020	George Suciu, Stefan Stefanescu, Marian Ceaparu, Cristian Beceanu	June 2020	Real-time communication and its advantages are discussed.	Streaming of audio/video data, change the IP addresses/ports with the other WebRTC clients.	Any parameter from the another LAN will not be able to access the end point's address.

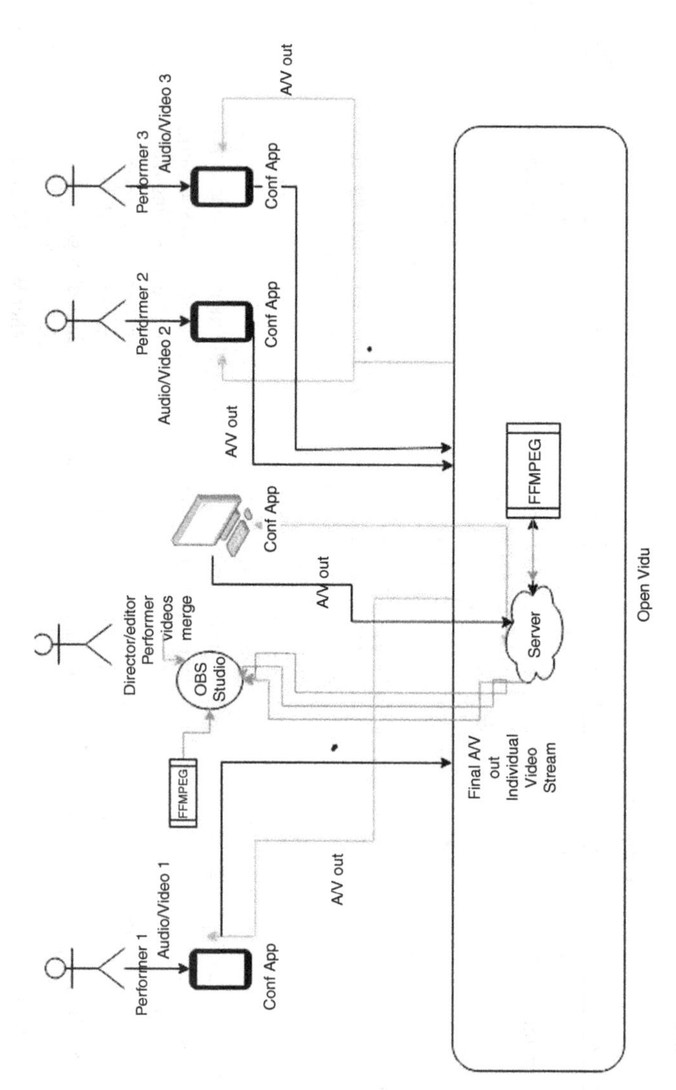

FIGURE 9.1 System architecture.

Details about OpenVidu, and AWS are given below.

- OpenVidu: A performer can connect through a cross platform application into the OpenVidu session. The director/host has control over the streaming, on the platform. They control the audio and video quality of the stream.
- AWS: The OpenVidu application is deployed using Amazon Web Services (AWS). With the help of this, we can provide our service to users with CloudFront technology.

9.5 TECHNOLOGY USED AND EXPERIMENTATION

Details of technologies illustrated in Figure 9.2 are explained below.

9.5.1 React (React, 2021, Telerik, 2021)

- React.js is a front end, open-source, JavaScript library for developing User Interfaces/UI components.
- It can serve as a base during the development of single page/mobile applications.
- Building React based applications requires additional libraries for routing, and state management.

Advantages of React:

- It is easy to learn and easy to use.
- Any individual having knowledge of Javascript can become familiar with React in a short time. This is very advantageous for newcomers who wish to start using this technology.
- It has reusable components. Every component controls the individual rendering and has its own inherent logic, which is reusable at any time in the future. Code reuse is an important trait, advantageous in making apps easy to build as also with easy future maintainability.
- It has great developer tools. It allows the inspection of the React component hierarchies in the virtual DOM. Virtual Dom is a React specific terminology. Individual components can be selected and examined and their current properties and states can be edited.

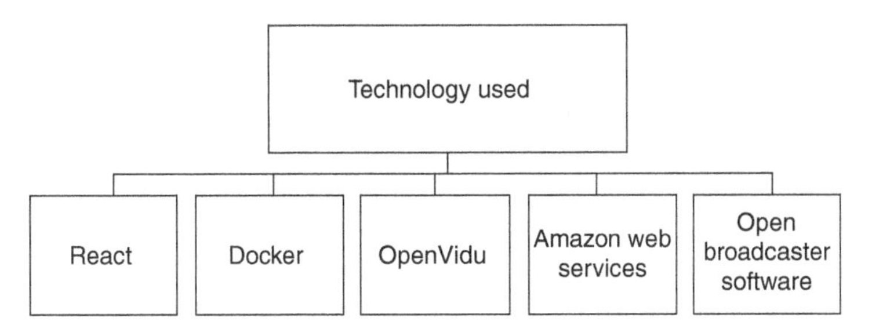

FIGURE 9.2 Overview of technology used.

9.5.2 Dockers

A collection of platform as a service products, Docker uses virtualization at OS level to ensure delivery of software in containers. These containers can be looked at as packages. Isolation is an important characteristic of these containers hence they bundle individual software, configuration files and libraries. These containers communicate with the other containers via distinct, pre-defined channels. A single kernel of operating system runs these containers, hence there is reported usage of comparatively lower resources in comparison to virtual machines (Docker, 2021; OpenVidu, 2021).

9.5.3 OpenVidu (OBS Studio 2021, Audio Video Streaming 2021, Live Streaming 2021, AWS 2021, OpenVidu 2021, AWS 2021, OBS 2021, AWS 2021)

OpenVidu is an open source WebRTC based application which facilitates internet based real time communication that helps to conduct video conferences and it includes many technologies that are useful for developing applications. Developing applications with the help of OpenVidu is easy and more efficient because it handles protocols that are part of WebRTC. It provides different APIs to make a WebRTC capable application that has an OpenVidu client and server side. Instead of overloading the system, it has new features which automatically let clusters grow or shrink according to the load on the CPU.

- OpenVidu Browser: This is the library used at client side. It is available for TypeScript and JavaScript. Using this browser, one is allowed the creation of video calls, join the users to these calls, receive and/or send files (audio, video) etc. It is through the OpenVidu browser that all the actions made available in OpenVidu are efficiently managed (Marhefk et al., 2020).
- OpenVidu Server: This is the application that successfully handles all the functionalities at the server side. The OpenVidu browser sends the operations to this Server. It is the Server that is technically responsible for establishing as well as managing the video calls. Individual users seldom have to explicitly use it (Apu et al., 2017). OpenVidu provides services such as: any combination of communications (one to one, for example) using WebRTC (Grozev et al., 2018), inter platform compatibility (for example, desktop apps, iOS, Android, Opera, Edge, Safari, Chrome, or Firefox) (Xu et al., 2012), or ease of use owing to simple (yet powerful and customizable APIs) (Wang et al., 2016).
- Message broadcasting: In OpenVidu, it is feasible to implement a chat with minimum (just a few) lines of code. Any type of communication that is based on text, and that occurs between the various users, is easily implemented in OpenVidu (Smiti et al., 2018).
- Recording: Record your video calls with complete freedom. OpenVidu provides predefined layouts, yet individuals have freedom to use customized layouts in their implementation of any task (Stefan et al., 2020).
- Screen sharing: In OpenVidu, clients are allowed to share screens for data exchange (Connolly et al., 2020).

- Audio and video filters: An important asset of OpenVidu is that it is the technology (WebRTC based) which allows users to apply audio and video filters in real time. To cite some examples, the detection of barcodes, the setup of a Chroma key background, the amplification of voices of clients, can be used. These and many more similar features are integrated in the APIs of OpenVidu (React, 2021).
- IP Cameras: IP cameras can be published in an OpenVidu session (Telerik, 2021).
- Implementation steps using OpenVidu:
  1. First of all, to make WebRTC based application run on different systems we install Dockers.
  2. Docker helps to download images by using a WebRTC server on Dockers and Virtual Machines.
  3. OpenVidu supports different platforms and according to that we can design our application.
  4. We used an application here, such as React.
  5. We used the JavaScript library so that we can design a user interface component.
  6. By installing the React library for OpenVidu we install the React version of OpenVidu.
  7. Here, changes related to application functions can be made by using JavaScript.
  8. Custom made changes can be made related to recoding, audio, video, or chat.
  9. For changes in these functions we use the JavaScript programming language.
  10. Then we run OpenVidu on a local host server
  11. Further development of the link is taken care of by AWS.

9.5.4 Amazon Web Services (AWS)

Amazon Web Services used in our system helps to provide on-demand platforms for cloud computing and APIs which can be administered on a prepaid basis. Cloud computing provides technical infrastructure, basic abstract, and building blocks for distributed computing. The AWS service offers us computing power, database storage and content delivery services. It also provides security for subscribers' systems.

9.5.4.1 Services Provided by AWS

- **Amazon Elastic Compute Cloud:** Amazon EC2 allows end users to have a virtual cluster. A part of Amazon.com. EC2 allows users to rent out virtual computers to run their individual applications. It equips users with full geographical control.
- **Amazon CloudFront:** We have used AWS CloudFront in our system as it is a content delivery web service with low latency and high data transfer speeds. CloudFront is a content delivery network (CDN) by AWS which provides a global distributed network of proxy servers. The proxy servers cache a variety

of content such as videos, or big data locally to the customers thereby enhancing the access speed of content download.

- **Amazon Simple Storage Service (Amazon S3):** We have used Amazon S3 in our project to store the OpenVidu.yaml file. Amazon S3 provides object storage through a web service interface. This service was used in the cloud formation step of deployment. Amazon S3 can be used for storage of internet applications, disaster recovery, backups, data lakes, data archives, hybrid storage, and analytics.
- **Amazon Identity and Access Management:** AWS identityand access management (IAM) is a web service that helps to securely control access to AWS resources. IAM is used to control who is authenticated (signed in) and authorized (has permissions) to use resources. This service was used in the instance creation steps of deployment. By using AWS the following was successfully accomplished.
- **AWS EC2 Instance Running:** An EC2 instance is created as it is a virtual server in Amazon's Elastic Compute Cloud (EC2) for running applications on the AWS infrastructure. The EC2 service allows us to run application programs in the computing environment. A demonstration of an instance is shown in Figure 9.3

The OpenVidu session is established after Dockerization of the OpenVidu React technology.

Once the stack is successfully created, an OpenVidu URL is generated from which connection to the OpenVidu server can take place and a session directly as illustrated in Figure 9.4

9.5.5 Open Broadcaster Software (OBS)

The base of the platform was created, with the help of OpenVidu. A platform was also needed to stream the live feed. Research indicated a platform, named Open Broadcaster Software (OBS). OBS is used to live stream any content from any random platform. OBS provides functionalities which makes streaming much easier and is very reliable. The process to Live Stream with OBS is very easy as the user only must use a URL and streaming key generated by the platform which is being used. OBS also provides some functionality that helps to improve the audio/video quality of the stream. By using such functionalities, the user can live stream content perfectly without any issues.

Open Studio is an open-source solution addressing game recording as well as live streaming. It is a free application which is expertly designed so as to enable multiple users to seamlessly work with various sources eventually, creating a hassle-free broadcast. It successfully enables the user to start recording from their webcam and microphone and also incorporates the surrounding videos, capturing the entire window/desired part of the screen, performs inserts in the games' footage (Telerik, 2021).

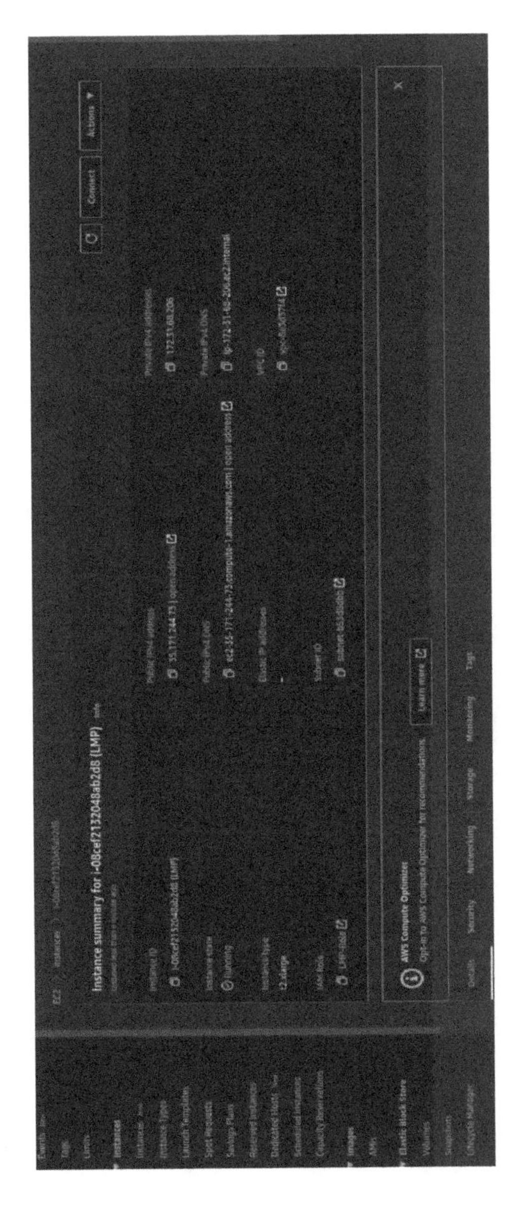

FIGURE 9.3 Running EC2 instance.

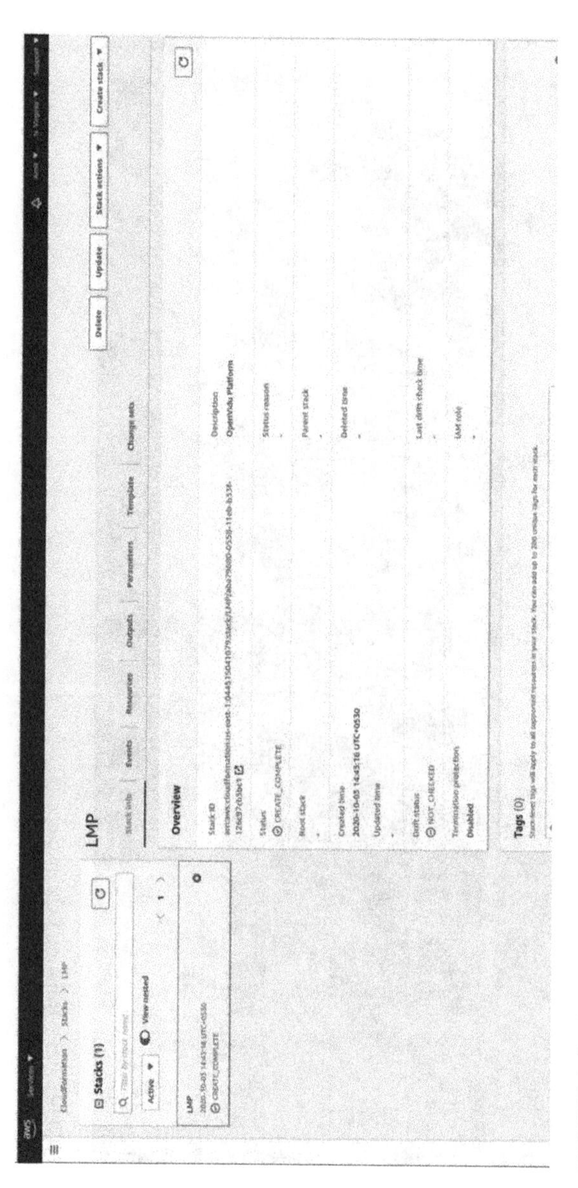

FIGURE 9.4 Screenshot of OpenVidu session.

9.6 APPLICATIONS

There are other applications apart from those covered in this work. Some of these are discussed below.

9.6.1 Healthcare

(Ignatowicz et al., 2019; Taylor et al., 2015; Raven et al., 2013; Stain et al., 2011; Wynn et al., 2012; Good et al., 2012; Edirippulige et al., 2013): For the past decade there has been a proliferation in the use of videoconferencing in the domain of healthcare. Notable advantages have been: seamless support to the patients living in rural/remote belts having a dearth of medical facilities, and efficiency with respect to time. As per previous research in this field, the following benefits are noted from patients' perspectives: reduction in the travel time to hospitals especially in the cases of medical emergencies/other societal issues and increased convenience for patients while seeking a consultation with doctors from the comfort of their own homes. This situation is also greatly advantageous both for elderly patients and for patients with long term illnesses.

Although there is much evidence of the effective use of videoconferencing in providing consultation between a patient and the respective healthcare practitioner, there are other societal issues to be considered, such as: the impact on expected outcomes, the cost factor, ethics, and challenges in implementing the systems for real world uses.

9.6.2 Education

(Lombardia et al., 2010): Research suggests that video conferencing has been successfully used as a learning tool among educators as well as the student community, worldwide. Advantages include promoting effective communication between the two (teachers and learners, peer groups, and so forth). This is especially true in the current days of the Covid pandemic where meeting in person is difficult/constrained. (Lombardia et al., 2010) addresses the various advantages as well as the issues in rendering higher education through videoconferencing. A detailed survey pertaining to videoconferencing of various types such as: interactive videoconferencing, desktop videoconferencing, web videoconferencing is also covered in depth. Results of the survey indicate probable differences in the learning outcome of learners/students. This is primarily due to lack of face-to-face communication. A summary of opportunities is presented below. Opportunities provided by desk videoconferencing include provision of an excellent collaborative environment for learners to share their ideas/ resources. This has an obvious collective benefit to all. Other advantages, when considering a larger geographical stratum, includes cultural exchange. Opportunities created by IVC also include direct interaction with the instructor which easily facilitates better performance and learning by students. Opportunities provided by WVC include the flexibility and freedom for teaching and learning to take place at the students' own pace. Challenges in desk videoconferencing include: providing the necessary technical/other support to learners to familiarize them and get them accustomed to the new learning paradigm. Learners may experience difficulties in dealing with linguistic variations, interruptions, learning in the new ways, due to hesitation/ reluctance to open up during discussions. Further research is required to examine the various cognitive behavioural factors of students' learning during their sessions through videoconferencing.

9.6.3 Driver Monitoring System

(Armfield et al., 2015): With the help of a smartphone that is mounted on the vehicle's dashboard, a cloud-based system for monitoring the drivers is presented in (Kashevnik et al., 2020). The authors claim it to be a first of its kind. Based on the history of driving, certain recommendations based on the current context are provided. Additionally, there is provision to recognize probable dangerous conditions. Based on how they are detected, the authors addressed two types of dangerous conditions as the driver being offline and the driver being online. Examples of online conditions to be detected include the presence of distractions, drowsiness, and so on. Examples of offline conditions to be detected include: drink driving, raised heart rate, aggressiveness while driving, mental stress, and so forth. Cloud resources were used appropriately for the successful detection of these conditions. Implementation methods included: mobile applications, statistics, web services, and analyses. The generation of reports was available for the supervisors. Another noteworthy advantage is that the current system blocks any use of other unrelated/unimportant applications, including messaging, in the driving sessions. In this way security is safeguarded. Despite the numerous advantages, the following limitations exist: discrepancies due to varying lighting conditions, unstable internet, use in areas where there is GPS unavailability, and so on.

9.7 NOVELTY

1. Our work is a novel experiment that supports polyphonic sound.
2. Users can watch on demand, but may also download and watch content needed at a later stage.
3. The application supports recording and chat functions.
4. The application is designed specifically keeping all of the key stakeholders in mind: artists, event managers, and so on.

9.8 RESEARCH CONTRIBUTION

- At the time of writing, due to Covid 19 conducting and attending meetings and concerts in person is not possible. Yet it is important for the ordinary person, and especially the theatre artists and musicians to continue their daily activities, for which an online mode has been a great advantage. WebRTC enables us to conduct meetings, events online and also to stream them. Users can watch the streams at home and at their convenience through the use of further AWS media services that will help them to download and watch at any time.
- One more point to be noted is that noise disturbances usually occur when musical notes are played in other applications. This has been overcome in the current work.

- The current work will be very useful for the seamless conduction of online concerts, for a geographically dispersed audience.
- Advantages of the current work include: cost effectiveness, reliability, efficiency, easy deployment, low latency, and enhanced user experience.

9.9 FUTURE WORK

- Based on the documentation supporting deployment of OpenVidu, the stream can be made available to users using CloudFront. Use of additional AWS services like Elemental Media could be employed to convert CloudFront in a future phase of this project.

9.10 CONCLUSION

- Recent years have seen a surge in widespread use of video conferencing applications. Challenges in its real time implementation include: seamless support for a large number of participants, the ability to efficiently deal with active changes pertaining to participant memberships, and most importantly, coping with a requirement for ever increasing bandwidth (Apu et al., 2017).
- In the current work, the system architecture and detailed experimentation and implementation of a novel good quality teleconferencing application using OpenVidu, has been presented. Use of the latter can help performers to showcase their talents across the world with the help of CloudFront technology where multiple users can join an online session where they can experience better video and audio quality facilitate the streaming of video on networks geographically. The system is cost effective, reliable, and efficient compared to other teleconferencing applications available on the market. This could be very useful in the current scenario. With the help of CloudFront technology, our application can easily be deployed in numerous regions globally with just a few clicks. Provision of reduced latency and richer experience for the customers at reduced cost is our objective. This will increase the speed and agility of the project. Our proof of concept described in this work will be very useful to the community.

9.11 ABBREVIATIONS

1. WebRTC: Web Real Time communication
2. AWS: Amazon Web Services
3. OBS: Open Broadcaster Software
4. RTSP: Real Time Streaming Protocol
5. HTTP: Hypertext Transfer Protocol
6. HLS: HTTP Live Stream
7. HTML: Hypertext Markup Language
8. RTC: Real Time Communication
9. DOM: Document Object Model
10. API: Application Programme Interface
11. CPU: Central Processing Unit

12 IP: Internet Protocol
13 EC2: Elastic Compute Cloud
14 CDN: Content Delivery Network
15 S3: Simple Storage Service
16 IAM: Identity and Access Management

ACKNOWLEDGEMENTS

The entire ownership of the current research (concept, design, source code, prototype) resides exclusively with the second author, who has provided the concept and problem statement relating to the current project. Amit Nene provided all technical support and guidance for the work, and has reviewed the manuscript. Suja Panicker served as college supervisor for the internship project described in this chapter. She coordinated and supervised the entire research paper writing process, guided the interns towards paper writing, proofread and reviewed/revised the contents alongwith actively contributing to the content. Ashish Hardas, Shraddha Kamble and Kaustubh Bhujbal served as interns at LastMinute Productions AU and implemented the current work as part of their academic internship project under the guidance and supervision of Company Supervisor, Amit Nene. Ashish Hardas was proactive in revising the draft under the guidance of Suja Panicker. All authors have contributed towards the writing of the paper.

REFERENCES

Al-Samarraie, H. 2019. A scoping review of videoconferencing systems in higher education learning paradigms, opportunities, and challenges. *International Review of Research in Open and Distributed Learning*, 20(3), pp. 121–140 doi.org/10.19173/irrodl.v20i4.4037

Armfield, N. R., Bradford, M., and Bradford, N. K. 2015. The clinical use of Skype: For which patients, with which problems and in which settings? A snapshot review of the literature. *International Journal of Medical Informatics*, 84, 10: 737–742.

Atwah, R., Iqbal, S., Shirmohammadi, S., and Javadtalab, A. 2015. A dynamic alpha congestion controller for WebRTC. *IEEE International Symposium on Multimedia (ISM)*, Miami, FL, USA, 2015, pp. 132–135, doi: 10.1109/ISM.2015.63.

Audio Video Streaming ffmpeg.org/ Accessed on 21/3/2021

AWS aws.amazon.com/codepipeline/ Accessed on 21/3/2021

AWS docs.aws.amazon.com/cloudfront/index.html/ Accessed on 21/3/2021

AWS aws.amazon.com/s3/ Accessed on 21/3/2021

AWS docs.aws.amazon.com/ec2/index.html/ Accessed on 21/3/2021

Drude, K. P. 2020. A commentary on telebehavioral health services adoption. *Clinical Psychology: Science and Practice*, 10.1111/cpsp.12325, 27(2). Wiley Online Library

Boris, G., George, P., Emil, I., and Thomas, N. 2018. Considerations for deploying a geographically distributed video conferencing system. *2018 IEEE 8th Annual Computing and Communication Workshop and Conference (CCWC)*, 2018, pp. 357–361, doi: 10.1109/CCWC.2018.8301726.

Cloud Services Continuum. 2008. et.cairene.net/2008/07/03/cloud- services-continuum/.

The Cloud Services Stack and Infrastructure. 2008. et.cairene.net/2008/07/28/the-cloud-services-stack-infrastructure/

Connolly, S. L., Miller, C. J., Lindsay, J. A., and Bauer, M. S. 2020. A systematic review of providers' attitudes toward telemental health via videoconferencing. *Clinical Psychology. Science and Practice*, 27, 2: 1–19.

Crandell, M. 2008. Defogging Cloud Computing: A Taxonomy. refresh.gigaom.com/2008/06/16/defogging-cloud-computing-a-taxonomy/.

De Moor, K., Arndt, S., Ammar, D., Voigt-Antons, J., Perkis, A., and Heegaard, P. E.. 2017. Exploring diverse measures for evaluating QoE in the context of WebRTC. *Ninth International Conference on Quality of Multimedia Experience (QoMEX)*, Erfurt, Germany, pp. 1–3, doi: 10.1109/QoMEX.2017.7965665.

Docker www.docker.com Accessed on 21/3/2021

Edan, N. M., Al-Sherbaz, A., and Turner, S. 2017. Design and evaluation of browser-to-browser video conferencing in WebRTC. *2017 Global Information Infrastructure and Networking Symposium (GIIS)*, Saint Pierre, France, pp. 75–78, doi: 10.1109/GIIS.2017.8169813.

Edirippulige, S., Levandovskaya, M., and Prishutova A. 2013. A qualitative study of the use of Skype for psychotherapy consultations in the Ukraine. *J Telemed Telecare*, 19, 7: 376–378.

Good, D. W., Lui, D. F., Leonard, M., et al. 2012. Skype: A tool for functional assessment in orthopaedic research. *J Telemed Telecare*, 18, 2: 94–98.

Hampel, R. and Stickler, U. 2005. New skills for new classrooms: Training tutors to teach languages online. *Computer Assisted Language Learning*, 18(4), 311–326. doi.org/10.1080/09588220500335455

Humer, E., Pieh, C., Kuska, M., Barke, A., Doering, B. K., Gossmann, K., Trnka, R., Meier, Z., Kascakova, N., Tavel, P., and Probst, T. 2020. Provision of Psychotherapy during the COVID-19 pandemic among Czech, German and Slovak psychotherapists. *International Journal of Environmental Research and Public Health*, 10.3390/ijerph17134811, 17(13).

Ignatowicz, A., Atherton, H., Bernstein, C. J., Bryce, C., Court, R., Sturt, J., and Griffiths, F. 2019. Internet videoconferencing for patient-clinician consultations in long-term conditions: A review of reviews and applications in line with guidelines and recommendations. *Digital Health*, 5, 2055207619845831. doi.org/10.1177/2055207619845831

Kashevnik, A., Lashkov, I., Ponomarev, A., Teslya, N., and Gurtov, A. 2020. Cloud-based driver monitoring system using smartphone. *IEEE Sensors Journal*, 20(12), pp. 6701–6671.

Ibn Zinnah Apu, K., Mahmud, N., Hasan, F., and Hossain Saga, S. 2017. P2P video conferencing system based on WebRTC. *International Conference on Electrical, Computer and Communication Engineering (ECCE)*, February 16–18, 2017, Cox's Bazar, Bangladesh.

Khalid, M. S., and Hossan, M. I. 2016. Usability evaluation of a video conferencing system in a university's classroom. In *Proceedings of the 19th International Conference on Computer and Information Technology (ICCIT)*, pp. 184–190. IEEE. doi.org/10.1109/ICCITECHN.2016.7860192

Krutka, D. G. and Carano, K. T. 2016. Videoconferencing for global citizenship education: Wise practices for social studies educators. *Journal of Social Studies Education Research*, *7*(2), 109–136. Retrieved from jsser.org/index.php/jsser/article/view/176/169

Live Streaming www.dacast.com/blog/how-to-broadcast-live-stream-using-ffmpeg/ Accessed on 21/3/2021

Flavio, L. and Roberto D. 2010. Secure virtualization for cloud computing. *Journal of Network and Computer Applications*, 34, 4: 1113–1122. doi:10.1016/j.jnca.2010.06.008

Marhefk, S., Lockhart, E., and Turner, D. 2020. Achieve research continuity during social distancing by rapidly implementing individual and group videoconferencing with participants: Key considerations, best practices, and protocols. *AIDS and Behavior*, 24:1983–1989

OBS Studio obsproject.com/ Accessed on 21/3/2021

OBS Studio reviews.financesonline.com/p/obs-studio/ Accessed on 21/3/2021

OpenVidu docs.openvidu.io/en/2.15.0/ Accessed on 21/3/2021

OpenVidu openvidu.io/tutorials/ Accessed on 21/3/2021

OpenVidu docs.openvidu.io/en/2.15.0/deployment/deploying-openvidu-apps/ Accessed on 21/3/2021

Raven, M., Butler, C., and Bywood, P. 2013. Video-based telehealth in Australian primary healthcare: Current use and future potential. *Australian Journal of Primary Health*, 19: 283–286.

React.js reactjs.org/ Accessed on 21/3/2021

Sabahi, F. 2012. Secure virtualization for cloud environment using hypervisor-based technology. *International Journal of Machine Learning and Computing*, 2(1): 39–45.

Smiti, P., Srivastava, S., and Rakesh, N. 2018. Video and audio streaming issues in multimedia application conference. *2018 8th International Conference on Cloud Computing, Data Science & Engineering (Confluence)*. IEEE, pp. 360–365, doi: 10.1109/CONFLUENCE.2018.8442823.

Sondak, N. E. and Sondak, E. M. 1995. Video conferencing: The next wave for international business communication. In *Proceedings of the Annual Conference on Languages and Communication for World Business and the Professions*, pp. 1–10. www.learntechlib.org/p/80886/

Stain, H. J., Payne, K., Thienel, R., et al. 2011. The feasibility of videoconferencing for neuropsychological assessments of rural youth experiencing early psychosis. *J Telemed Telecare*, 17: 328–331.

Stefan, G. S., Beceanu, S. C., and Ceaparu, M. 2020. WebRTC role in real-time communication and video conferencing conference. *2020 Global Internet of Things Summit (GioTS) IEEE*, Dublin, Ireland.

Taylor, A., Morris, G., Pech, J., Rechter, S., Carati, C., and Kidd M. R. 2015. Home telehealth video conferencing: Perceptions performance. *JMIR Mhealth Uhealth*, 3(3): e90.

Telerik www.telerik.com Accessed on 21/3/2021

Wang, J., Xu, W., and Wang, J. 2016. A study of live video streaming system for mobile devices. *International Conference on Computer Communication and the Internet (ICCCI)*. IEEE, Wuhan, China.

Wynn, R., Bergvik, S., Pettersen, G., et al. 2012. Clinicians' experiences with videoconferencing in psychiatry. *Stud Health Technol Inform*, 180: 1218–1220.

Xu, H., Chen, Z., and Cao J. 2012. Live streaming with content centric networking. *The Third National Conference on Networking and Distributed Computing*. IEEE, Hangzhou, China.

Youseff, L., Butrico, M., and Dilma D. S. 2008. Toward a unified ontology of cloud computing. *Grid Computing Environments Workshop*. IEEE, Austin, TX, USA, 2008, pp. 1–10, doi: 10.1109/GCE.2008.4738443.

10 Agriculture 5.0 in India

Opportunities and Challenges of Technology Adoption

Rajesh Tiwari, Khem Chand, Arvind Bhatt, Bimal Anjum, and Thirunavukkarasu K.

CONTENTS

DOI: 10.1201/9781003138037-10

10.1 INTRODUCTION

This section introduction discusses the need for Agriculture 5.0 in India, scope for artificial intelligence, mobile applications, drones, blockchain in agriculture, and drip irrigation.

10.2 AGRICULTURE: PROBLEMS AND SOLUTIONS

10.2.1 Agriculture in India

Agriculture accounts for 6.4% of economic production of the world. Agriculture employs 40% of the global workforce contributing US $ 5 trillion in worldwide production (Tomu, 2020). The large proportion of India depends on agriculture. 800 million people in India depend on agriculture contributing 16% in GDP (Prakash and Parija, 2019). Poor mechanisation leads to poor productivity in agriculture in India. The level of mechanisation in agriculture is 40% in India, as compared to 59.5% in China and 75% in Brazil (Pib, 2020).

The poor productivity of Indian farmers is a cause of concern. According to National Institution for Transforming India's (NITI Aayog) Vice Chairman, 80% of poor people in India depend on farming (Huffpost, 2017). The yield of paddy in India is 3500 kg per hectare as compared to 7000 kg in China, 10,000 kg in Australia. The productivity of wheat in India is 3000 kg as compared to 5000 kg in China (Sangal, 2018). Prime Minister Mr. Narendra Modi has made a commitment to double farmer income by 2022. Achieving a milestone of a US $ 5 trillion economy is difficult without income growth of the approximately 800 million people who depend on farming. $6.7 billion has been invested in agriculture technology in the last 5 years. The market for smart farming is expected to increase by 13.27% compound annual growth rate to reach $ 13 billion by 2021 (DCunha, 2018). The adoption of technology in agriculture is constrained by lack of credit, lack of awareness, small farm size, poor supply of complementary inputs, risk aversion, and insufficient human capital (Feder et al., 1985).

The success of Imperial Tobacco Company's (ITC) e-Choupal, is a beginning in technology adoption among farmers in India. Launched in 2000, ITC's e-Choupal is the world's largest rural internet initiative for farmers. ITC's e-Choupal is empowering 4 million farmers from over 35000 villages in 10 states connected through 6100 kiosks (itcportal, 2020). E-choupal has been recognised for its role in empowering farmers using technology with awards at national and international level. The Government, led by Mr. Narendra Modi has enhanced confidence with an enabling environment (Rana and Tiwari, 2014).

Precision farming is concerned with increasing production with fewer resources, lesser cost, and by managing variations in the field (CEMA, 2018). Digital farming combines precision farming with networks and data management (CEMA, 2018). Data analytics can be used to automate processes in agriculture. Digital farming relies on connectivity between machine and machine, machine and cloud and cloud and cloud. Agriculture 4.0 also referred to as digital farming or smart farming, focussed on precision by combining telematics and data management. Agriculture 5.0 involves use of precision farming along with robots and artificial intelligence (Khan and Kannapiran, 2019; Rubio and Mas, 2020). Technology in agriculture can have a transformation effect by solving issues related to quantity, quality, storage and, distribution (Chakravarty, 2018). Technology in rural areas and farming positively influences the empowerment of women (Anjum and Tiwari, 2012a). Rural households are facing employment problems due to poor productivity in agriculture caused by lack of access to advanced technology. Unemployment and poor skills adversely affects economic growth (Khem et al., 2017).

Industrialisation will lead to a decline in arable land in India. The western states of India: Maharashtra, Gujarat, Rajasthan, and Madhya Pradesh are estimated to have 65% urban population by 2030 (Maiervidorno, 2020). With a growing population and declining arable land, technology is the feasible solution to provide sustainable supply of food and improve the productivity of agricultural processes. Agricultural land is gradually being converted into non agricultural uses due to the pressure of urbanisation. Technology can increase productivity and make up for the loss of agricultural land. Robotics has reduced the operating costs and improved productivity in several countries (Reddy et al., 2016). However, affordability for small farmers needs to be worked out to make it feasible in India. Financial engineering is needed to make the cost of robotics affordable for small and marginal farmers. Empowerment of rural households, better governance, empowerment of women and access to better quality of life including access to education (Anjum and Tiwari, 2012a; Anjum and Tiwari, 2012b; Khan et al., 2021) can be enhanced by technology in agriculture.

Water scarcity is a big challenge for India. 78% of water is used for agriculture in India (Reddy, 2020). Technology can enhance the efficiency of water usage for agriculture. Technology can bring down operational costs by reducing usage of fertilisers and pesticides. Technology positively impacts the environment with reduced movement of chemicals to rivers and ground water (Reddy, 2020). Drip and sprinkler irrigation systems can reduce water consumption by 70% and enhance yields of crops by 20 to 90% (Reddy, 2020). Green housing cultivation can improve output and can also be used for rainwater harvesting. In territories with 400 mm yearly rainfall falling on polyhouse structures, that is 70,000 litres, even at 80% capacity 56000 litres water can be harvested (Reddy, 2020).

Crop management in India has been through traditional methods. Technology can bring a scientific perspective in crop selection by analysing demand, pricing, and weather conditions Robotics offers the opportunity of the automated grading and sorting of crops (Chakravarty, 2018). Supply chain inefficiencies leads to a huge loss of agricultural output. Technology adoption can bring efficiency and reduce the wastage. ITC's e-Choupal had a ripple effect on the food grain management system of the public sector (Seth and Ganguly, 2020).

This chapter explores disruptive technologies for the agriculture sector. The crucial issues impacting farmers with regard to access, understanding, affordability, and implementation of technological interventions have been discussed. The chapter makes a contribution to operationalising Agriculture 5.0 by proposing a model to connect with farmers to convert them into new age farmers with technology as a tool for enhancing efficiency and economic outcomes.

10.2.2 Artificial Intelligence

The use of artificial intelligence provides opportunities to take informed decision on crop yields, pest management, crop disease prediction, weather forecast, soil health and ground level indicators. CropIn Technologies in collaboration with the World Bank is working in Bihar and Madhya Pradesh to provide data on historical crop yields that can be used to forecast future yields using machine learning. Remote sensing is used to track crops. The Weather Company, an initiative of IBM provides data on temperature and soil moisture so that farmers can make a good call on the right time to irrigate. The weather unit along with Agro Star is working to develop crop disease forecast algorithms. IBM has signed a statement of intent with the Ministry of Agriculture, Government of India to use artificial intelligence for providing technology solutions in the field of agriculture (Mendonca, 2019). Agnext is an IIT-Kharagpur incubated company working on artificial intelligence based technology solutions for quality estimation in the agriculture sector. Agnext has been awarded best Agri-tech company in Asia by Rabobank SustainableAg. Agnext provides artificial intelligence (AI) based image analytics, spectral analytics, and sensor analytics (Agnext, 2020).

10.2.3 Mobile Applications

Smart phone penetration in rural areas provides an opportunity to connect farmers with technology through mobile applications. A government initiative of the electronic national agricultural market (e-NAM) provides an electronic trading platform for agricultural commodities. The Punjab Government has launched the Punjab Remote Sensing Centre (PRSC) to connect with farmers. PRSC offers services through i-khet machine, e-PeHal and e-Prevent (Udas, 2020). Mahindra and Mahindra had launched a tractor rental app for farmers, called Trringo. Trringo has over 1.5 lakh (1 lakh = 100 000 in the Indian numbering system) registered farmers with 2.5 lakh hours of farm mechanisation since its launch in 2016. Trringo won the IDC digital transformation award (Trringo, 2020). Trringo lowers the cost to farmers as they only pay for the actual usage and save the capital investment. Tata trusts with the support of Tata Consultancy Services (TCS) has developed mKrishi mobile for technology enabled last mile connectivity and an advisory role for farmers. The mKrishi app provides advisory services on weather updates, pest control, best practices through SMS, IVR and calls on mobiles. The mKrishi app has 4 lakh users from Punjab, Gujarat, Tamil Nadu and Maharashtra getting expert consultancy on nine crops (Tatatrusts, 2020).

10.2.4 Drones

Drones provide an opportunity to collect real time data using sensors for soil conditions, crop growth, dry regions, pest and crop diseases, and spraying. Dynamic remotely operated navigation equipment (DRONE) also referred to as an unmanned aerial vehicle (UAV) provides the opportunity to remotely monitor crops, soil conditions, weather conditions, to make crop loss assessment, and for various allied activities in agriculture (Mohapatra, 2016). DRONES can also be used for planting seeds and trees, weed identification, crop spraying, protection from damage of crops by animals, detection of diseased animals, and scheduling of the irrigation of crops. Drones can be used for reforestation and afforestation (Rani et al., 2019). 145 agriculture drone companies are working globally (Traxcn, 2020). In India, more than 40 start-ups are working in the field of drones (Rawat, 2020). According to Jose Luis Fernandez, FAO representative in the Philippines,

> The adoption of modern technologies in agriculture, such as the use of drones or unmanned aerial vehicles (UAVs), can significantly enhance risk and damage assessments, and revolutionize the way we prepare for and respond to disasters that affect the livelihoods of vulnerable farmers and fishers and the country's food security
>
> (FAO, 2020)

Drone cans make assessments of 600 hectares in a single day (FAO, 2020). This can speed up damage assessment and quantification. TartanSense is a Bangalore based, UAV based start-up, established in 2015. TartanSense uses near infra red sensors to detect healthy plants. It can analyse aerial images of 700 hectares in a single day. At the moment, UAVs are used for cutting of trees over large areas (Traxcn, 2020). 1 Martian Way Corporation is a Mumbai based firm providing artificial intelligence (AI) based products for drones. Cron Systems offers multi sensor scanners for UAVs. Aarav Unmanned Systems (AUS) is a leader in commercial drones in India. AUS was incubated by IIT Kanpur in 2013 (AUS, 2020). Aerial Photo provides aerial imagery. Drones Tech Lab manufactures drones for pesticides, crop monitoring, and surveillance (Shrivastava, 2020).

10.2.5 Blockchain

Blockchain is a distributed ledger of transactions and digital events shared among the participating parties. The data is kept on a network of computers instead of a physical ledger or single database (Patel et al., 2017). Transactions are approved through consensus and secured through cryptography. Transactions are made at a peer to peer level, without any third party control. Blocks store the transactions in an encrypted manner, which cannot be tempered. The 'chain' in blockchain refers to the connection of various blocks together, thus creating a database which has information relating to every stage of the transaction. Once an entry is made, it cannot be edited or deleted by any member. In 2008, an entity, called Satoshi Nakamoto developed virtual currency, bitcoin using blockchain (Nakamoto, 2008). Blockchain has moved beyond virtual currency use to many sectors. According to Ginni Rometty, CEO, IBM,

> Anything that we can conceive of as a supply chain, blockchain can vastly improve its efficiency- it doesn't matter if its people, numbers, data, money.
>
> (Leiker 2018)

NITI Aayog has recommended establishing India's own blockchain, named IndiaChain (2020). Eka software has built a blockchain platform in association with the Coffee Board of India for commodity management by farmers. State governments of Tamil Nadu, Andhra Pradesh and Kerala are working on the blockchain platform for shrimp and cashew nuts. Tata Trusts is working with the Marine Product Export Development Authority for institutionalising blockchain for the fisheries sector (Mendonca, 2019).

Blockchain innovations in the field of agriculture are estimated to grow at a compounded annual growth rate (CAG) of 47.8% from US $ 41.2 million in 2017 to US $ 430 million by 2023 (Startup Insight, 2020). Blockchain benefits the agriculture sector in crop insurance, the optimisation of the supply chain, traceability, and transactions (Startup Insight, 2020).

10.2.6 Drip Irrigation

The use of drip irrigation will reduce consumption of water and improve productivity. Irrigation in India consumes 80% of water in India due to the tradition Flood Method of Irrigation (FMI). Usage of water in India is 2 to 3 times more in comparison to China and Brazil. Drip irrigation uses 30 to 70% less water and increases crop productivity by 30 to 90%. Drip irrigation reduces electricity consumption by 30% and fertiliser consumption by 28%. Pradhan Mantri Krishi Sinchayee Yojana, launched in 2015 to promote micro irrigation, has led to an increase in the area under micro irrigation. The area under micro irrigation in 1985–1986 was 1500 hectares, which increased to 4.24 million hectares in 2017. India has lot of potential for drip irrigation as only 4% of area is under drip irrigation according to the data of 2016–17 (Narayanamoorthy, 2019).

10.3 LEVERAGING DIGITAL INFRASTRUCTURE

As presented in the following table (see Table 10.1), India has the largest pool of biometric identities in the world with 1.2 billion aadhaar cards (unique identification cards). India's mobile teledensity is 90.52%. India has 514.56 million rural subscribers of wireless phones (TRAI, 2020). Internet subscribers have grown at a compound annual growth rate (CAGR) of 45.74% during the financial years 2006 to 2019. Total internet subscribers are 665.31 million (IBEF, 2019).

Existing digital infrastructure will support blockchain enabled services for farmers in India. The political will of the existing NDA government led by Prime Minister Mr. Narendra Modi will further boost technology start-ups by developing a regulatory framework. The 5-I model proposed by Mr. Narendra Modi in the G20 summit on digital economy and artificial intelligence, focuses on Innovation, Inclusiveness, Indigenisation, Investment in Infrastructure and International Cooperation (the five I's) (Chaudhury, 2019). The vision of the Modi Government is to leverage technology for social benefits. A direct benefit transfer scheme eliminated the middleman and benefitted 120 million people (Chaudhury, 2019).

India has a talent for digital initiatives. It was found that among a five emerging markets study comparing India, Nigeria, Mexico, Poland and the Phillippines, the best performer in terms of talent was India. India was also best among the five nations in terms of attracting entrepreneurs and investors, and best at innovation. Conversely, India's weaknesses lie in public health, environment, and inclusion

TABLE 10.1
Digital Infrastructure of India

Aadhaar	1.2 bn biometric records
Unified Payment Interace	1.3bn transactions (December 2019)
mobile Subscribers	1173.75 million
Urban Mobile Subscribers	659.18 million
Rural Subscribers	514.56 Million
Urban teledensity	156.18
Rural Teledensity	57.28
Internet Subscribers	665.31 million
Goods and Services Tax Network	More than 400 million returns, More than 800 million invoices
Prime Minister Jan Aarogya Yojana	119 million ecards issued, 8 million hospital admissions, 500 million beneficiaries covered

Source: NITI Aayog 2020, India Brand Equity Foundation 2019.

(Chakravorti and Chaturedi, 2019). Demonetisation was a bold move to promote digital payments. Mr. Narendra Modi has emerged as the prominent digital influencer in India.

10.4 FINTECH

The entry of big players in the Indian fintech sector has intensified competition. Google Pay has become the largest unified payment Interface (UPI) touching 40700 crore INR (1 crore = ten million in the Indian numbering system) transactions in April 2019 (Kapoor and Usmani, 2019). Larger players will not only create barriers to entry for the new players but also threaten the existence of small players. Existing players can explore diversifying into other financial products in order to sustain their operations. Collaborations between banks and fintech will benefit both the entities. The start-up culture is gaining popularity as the number of start-ups has increased from 1800 in 2010 to 40,000 in 2019 (Kapoor and Usmani, 2019). The India start-up ecosystem has successfully produced 25 Unicorns (privately held start-up companies) up until 2019.

10.5 UBERISATION

Technology adoption in India is facing problems due to lack of capital among small and marginal farmers. Uberisation of technology-enabled services would bring down the cost per user and provide the benefits of the latest technology to farmers at the bottom of the pyramid. The Trringo tractor rental app is an initiative for small and marginal farmers to rent a tractor for farm use without owning the tractor (Trringo, 2020). Poor utilisation of resources threatens financial viability (Anjum and Tiwari, 2012b). Uberisation supported by an enabling framework from government, banks,

NBFC, or development banking institutions, can bridge the digital divide between rich farmers and poor farmers. Uberisation can make robotics and artificial intelligence products feasible for small and marginal farmers.

10.6 OPPORTUNITIES FROM BLOCKCHAIN IN AGRICULTURE

10.6.1 Land Records

Lack of land records affects marginal and tenant farmers, as they are not able to access credit. Though computerisation has been initiated by several state governments, only in 39% of villages has spatial data been verified up until 2017 (Padmanabhan, 2018). The use of blockchain in providing digital identities will resolve this basic problem faced by marginal farmers. Tenant farmers account for 80% of all farmer suicides (Raju, 2019). Tenant farmers are not eligible for Pradhan Mantri Fasal Bima Yojana (a government sponsored crop insurance scheme) due to lack of land records. After liberalisation in the period 2003 to 2013, tenancy increased 10.4% as per the 70th round of the National Sample Survey Office (NSSO) report. Andhra Pradesh with 35.7%, Bihar with 22.7%, Haryana with 14.8%, Odisha with 16.9%, Tamil Nadu with 13.5% and West Bengal with 14.7% were above the national average of 10.4% tenancy (Raju, 2019).

Land records are not completed regularly by state authorities. A land survey in the Telangana region (previously in Andhra Pradesh) was done during the regime of Nizam in 1932–36 (NITI Aayog, 2020). Land disputes account for two third of all pending cases in Indian courts (World Bank, 2007) and litigation involving land disputes take about twenty years to resolve (Debroy and Jain, 2017). NITI Aayog has developed a prototype based on studies done in the Union Territory of Chandigarh.

States with high land reform activities had better outcomes for households in terms of income growth, assets, and educational accomplishments (World Bank, 2007). Raju (2019) recommended setting up of the Tenant Farmers Development Fund for refinancing, crop insurance, calamity relief, and skill development.

10.6.2 Crop Insurance

Crop loans are mandatory to get crop insurance. The majority of farmers are not able to access crop insurance due to lack of documents. Blockchain would provide digital identity to the land, enabling farmers to access crop insurance. The evaluation of crops is done by crop cutting experiments (CCE). This is prone to human error and manipulation. Delay in crop yield and loss evaluation through CCE by state governments is the major reason for delay in payment of insurance claims to farmers. The use of blockchain will provide real time data about weather conditions and the evaluation process will be streamlined to provide quick reports for timely claim settlements of crop insurance. Gramcover focuses on offering digital insurance for low and middle income farmers (Kapoor and Usmani, 2019).

Worldcovr provides crop insurance to small farmers in Africa. It uses satellites to assess rainfall and make payments automatically based on scientific assessments (Worldcovr, 2020). Worldcover has raised US $ 6 million in series A funding for expansion in emerging markets, including India. Worldcovr has provided services to more than 30,000 farmers in Africa. Farmers connect using a mobile app, they make payments using mobile phones and claims are filed by farmers using mobiles. The company uses scientific evaluation and after certification, insurance money is paid to farmers through their mobiles (Bright, 2019).

Etherisc provides a free open source, open access platform for insurance using blockchain. Farmers can create customised insurance products as per their risk assessment and requirement. Ethreisc is focussed on decentralising the insurance offering. It uses a two fold approach. Firstly, it is a non-profit making entity, namely, the Decentralized Insurance Foundation, and secondly, it comprises of multiple commercial entities. Etherisc launched crop insurance in Sri Lanka in July 2019 in association with Aon, Oxfam (Etherisc, 2020). Technology adoption has the potential to enhance the benefits of crop insurance in India (Tiwari et al., 2020).

10.6.3 Small Farms

According to the agriculture census of 2015–16, published in 2018, small and marginal farmers owning less than 2 hectares of land comprise 86.2% of all farmers in India, but own only 47.3% of the crop area. Table 10.2 shows the categorisation of farmers.

Bihar is a high population state with the lowest average size of holdings as shown in Figure 10.1. The national average holding is 1.08 hectare. Uttar Pradesh (0.73), West Bengal (0.76) and Andhra Pradesh (0.94) have land holdings below the national average. Nagaland has the largest average land holdings at 5 hectares followed by Punjab with 3.63 hectare. The average size of a farm is less than 1 hectare in 12 states (Ministry of Agriculture & Farmers Welfare, 2019).

126 million small and marginal farmers own 74.4 million hectares of land, averaging 0.6 hectares each. Such a small land holding is not enough to financially sustain the families of farmers (Bera, 2018). In 2013 marginal farmers earned less than 5500 INR a month from farming and non-farming activities (Chakravorty, Chandrasekhar, and Naraparaju, 2016). The numbers of such farmers were 100 million according to the 2015–16 data (Padmanabhan, 2018).

Use of blockchain to provide a digital identity can be used to pool land. Pooling of land will make it easy to develop a land leasing market. China has made land reforms to manage the small farm problem. There is co-ownership of land between village

TABLE 10.2
Category of Farmers in India Based on Farm Size

Type	Size (hectare)
Marginal	Below 1.00
Small	1.00–2.00
Semi- Medium	2.00–4.00
Medium	4.00–10.00
Large	More than 10.00

Source: Ministry of Agriculture & Farmers Welfare, 2019

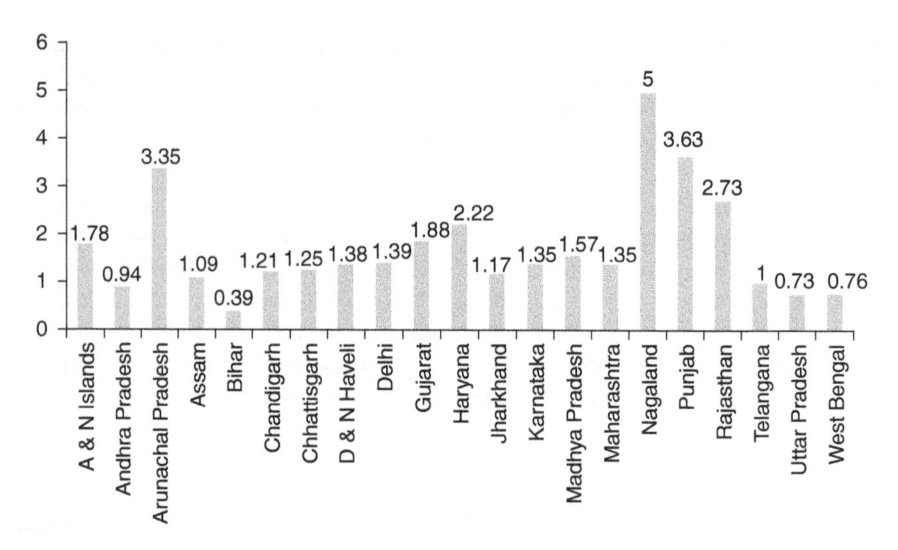

FIGURE 10.1 Average size of land holdings (hectare) as per agriculture census 2015–16.

and household. Land ownership remains at the village level and land use rights are with household. This facilitates land use rights between households. The land tenure system which commenced with a planned ten year lifespan, has been extended to 30 years duration. The land rental market benefitted small farmers in China (Huang, Wang, and Qiu, 2012).

Demeter is a blockchain based platform to rent micro farms anywhere in the world without any middlemen. Users and growers transact directly. Farmers get better returns and users select micro farm, practice and produce. Demeter refers to the approach as ag4.0 or growfunding (Demeter, 2020). It will assist organic farming.

10.6.4 Certifying Organic Farming

Organic foods exports from India have increased 39% from $370 million in 2016–17 to $ 515 million in 2017–18 (Sally, 2018). Few states have taken organic farming as top priority. Madhya Pradesh leads with the largest covered area committed to organic farming. Rajasthan, Maharashtra and Uttar Pradesh are other leading states. Sikkim has converted all cultivable land for organic farming (Sally, 2018). Organic farming has immense potential as it just accounts for 3% of agricultural export (Mukherjee, 2018).

Organic exports face the challenges of certifications from multiple agencies. Participatory guarantee systems (PSG) and third party agencies provide certifications. PSG certifications are not accepted by major markets. Third party certificates are costly. The use of blockchain for updating information on the lifecycle of crops will enhance traceability, peer to peer certifications, integrating third party certifiers into the PSG mechanism will enhance acceptance of organic farming products globally (NITI Aayog, 2020).

10.6.5 Credit

Without land records crop loans are not provided. Tenant farmers depend on money lenders who are not considered in the farm loan waiver by state governments (Raju, 2019). WhatsApp will enable users to transact in Bitcoin and Litecoin (Ganguly 2019). According to PriceWaterhouseCoopers (PWC), 77% of fintech companies are expected to adopt blockchain in their operations by 2020 (PWC, 2020). Technology has enhanced financial inclusion with private sector participation (Anjum and Tiwari, 2012c).

10.6.6 Supply Chain

Supply chain for agriculture is dominated by intermediaries. Intermediaries exploit farmers. Apte and Petrovsky (2016) argued that blockchain can remove the distortions in supply chains. Since blockchain is a peer to peer mechanism farmers can trade directly with customers and remove all middlemen. According to Report Linker, blockchain in the food and agriculture supply chain will increase from US$60.8 million in 2018 to US $ 429.7 million in 2023 (Hertz, 2019).

Agrichain is a company dealing in the agricultural supply chain based on a blockchain platform. It is used by farmers and industry participants from Australia and Asia (Agrichain, 2020). Agrichain connects various stakeholders in a secure, transparent manner, cutting cost and improving efficiency. It is free for farmers and a mobile app based service provides an easy way to manage stock and logistics (Agrichain, 2020).

Ripe is a blockchain based supply chain for food. Ripe provides a platform to access the origin of the product, the journey of the product and quality in a reliable and transparent manner. Partners can access real time data on the mobile app. Everything is uploaded on blockchain to ensure accessibility at all times. Partner includes producers, distributors, food retailers and restaurants (Ripe, 2020).

Blockchain will also check wastage of food. Globally one third of the food, about 1.3 billion tonnes, costing $940 billion is wasted or lost (Welvaert, 2020). In India, 67 million tonnes, worth 92000 crore INR of food is lost every year (Haq, 2019). Blockchain will connect growers with users and the wastage can be minimised.

10.7 CHALLENGES TO THE ADOPTION OF TECHNOLOGY

The adoption of technology in a developing, diverse and democratic country cannot be studied as a uniform challenge. The diversity of the country necessitates region specific policies for the adoption of technology. Willingness to pay is determined by behavioural, socio-demographic, technological, institutional, and biophysical factors (Solomon et al., 2019). Joblaew et al. (2020) found that technology adoption by farmers is impacted by gender, rice farming experience, attendance of rice production technology training, agricultural extension officers, problems relating to the project, and household income. Non adopters lack skills and financial resources (Christensen and Raynor, 2003). Collaboration with industry enhances opportunities for skill development (Tiwari and Anjum, 2014).

A regulatory framework in India has not been developed to legalise and regulate blockchain based currency and other services. Blockchain is still in its infancy in India. Absence of an enabling policy framework hinders the growth of blockchain based start-ups. Financial discipline and dedication in the team is useful in managing a challenging operating environment (Choudhuri et al., 2015). Technology, autonomy, and better governance enhances outcomes (Sharma et al., 2016; Khan et al., 2020).

Farmers need to be skilled to safeguard themselves against the misuse of technology. Labour market reforms and the social security framework has a positive impact on inclusive growth (Anjum and Tiwari, 2012d). Growing cybercrime is a cause for concern. State governments are at loggerheads with central governments on various issues. Lack of synergy and coordination with regards to policy for digital initiatives may hinder the growth of blockchain start-ups. Myopic vision of politicians to keep farmers poor, so that they can keep promising farm loan waivers, and other freebies to get votes in every election, might restrict the growth of blockchain start-ups in India. Sanctity of contracts, rigid labour laws, complex land record administration, and lack of political will are challenges for blockchain adoption in India (Mulraj, 2019). Shortages of blockchain experts can restrict growth of blockchain in India. The cost of blockchain experts is higher than data scientists and software developers. It is difficult to build trust in blockchain technology due to conflicting views and political will in India across different states for blockchain.

Exits have been a sore point in the start-up ecosystem in India. M R Rangaswami, venture capitalist from Silicon Valley, commented that liquidity and exits are not as per expectations (Kapoor and Usmani, 2019).

10.7.1 Communication with Farmers

The communication failure is a major setback for technology adoption in India. Policy makers and administrators are not able to connect with farmers.

10.7.2 The Subsidy Model

The subsidy model has led to suppression of innovative solutions, and distorted the market towards to obtaining subsidies rather than efficiency for farmers. A political approach to retain the vote share by appeasing farmers has led to continuation of subsidy policies, without any significant sustainable fruitful results for small farmers. The focus of the firms is more on subsidy share rather than farmer welfare.

10.7.3 Fragmented Farm Holdings

The small size of farm holdings increases the cost of transaction

and firms tend to avoid small farmers. Small farmers haven't got any collective organisations/associations to explore the adoption of technology on a sharing basis as they are so fragmented.

10.7.4 Poor Value Addition

Policies for farmers have only focussed on the subsidy and procurement at support prices. The price realisations have remained poor due to intermediaries and this has restricted the financial capability of farmers. No institutional effort has been made to empower farmers to use their farm produce to develop value added products to enhance profitability.

10.8 CARE MODEL FOR AGRICULTURE 5.0 IN INDIA

Agriculture 5.0 needs a radical shift in the approach towards farmers. Instead of considering them as a mere vote bank, they need to be considered as a potential source of value creation for a vibrant and sustainable economic growth. The incremental approach needs to be changed with disruption to redefine traditional approaches with technology driven scientific approaches.

Free demonstrations and usage rights for a season or one year will not only boost the confidence of the farmers but also bring them closer to technology. Technology adoption by small farmers is impacted positively by demonstration trials and free access to technology for at least the initial phase of using new technology (Yigezu et al., 2018; Xiang et al., 2021). New technology should be compatible with local needs and competitively priced. Extension services should be provided to farmers by the government and established institutions (Sirisunyaluck, Singsin, and Kanjina, 2020).

The proposed Care model for enhancing adoption of technology in Indian agriculture is based on adding a human touch by caring for farmers in a holistic manner.

1. Care to communicate: Simplify the communication to farmers so that they can understand and use the scientific inputs. Communication should be in local language and should use a (easy to use) mobile based application.
2. Care to convince: The small farmers should be convinced about the benefits of technology adoption by demonstrating the application on their farms.
3. Care to give concessions for first time use: The technology products and services should be provided free of charge for one season or one year to farmers so that they are able to witness the benefits in their lives.
4. Care to give credit: The credit facilities should address the unique needs of the region and should be tailor made to motivate farmers to adopt the technology most beneficial for their area.
5. Care for consultancy and training: Consultancy and training for farmers should be institutionalised by a collaborative framework.
6. Care for cluster: Clusters with unique needs and opportunities should be identified. The policies should be tailor made to suite the cluster. The special economic zone (SEZ) model should be used to give focussed attention to farmers within a particular cluster. The export potential and the food processing of such clusters should be enhanced.

7. Care for cooperative farming: Small farm size is a constraint for the adoption of technology. Cooperatives should be formed by merging the small farms so that farmers retain the ownership and total cumulative farm size is used for technology products and services to bring down cost of acquisition and operation. Amul is a cooperative milk federation, which has changed the lives of millions of farmers engaged in milk production activities.
8. Care for extension services: The institutions should provide care by way of extension services. New technology has many hurdles in terms of operations and maintenance. To make it viable and to adopt new technology for farmers, extension services must be strengthened to provide much needed care.
9. Care for contract farming: Middle and large sized farm holders should explore contract farming to leverage benefits of technology. Such farmers can help in technology diffusion to the small farmers by mentoring.
10. Care for profits by organic farming: Organic farming can create value by increasing the monetary realisation of the output. Organic products provide more revenues as compared to traditional products.
11. Care to create market to deal with end users: Middlemen grab most of the profits of farm produce and farmers remain poor. To enhance better price recovery, it is vital to create a market to deal with end users directly. Cooperatives should be formed and the cooperative should deal with end users directly. Collaboration with online portals can provide an opportunity to deal with end users and get price realisations for farm produce.
12. Care to consolidate financial offerings: The financial offerings to farmers should be merged together to provide an umbrella cover to fulfil all needs, beginning with seeds, fertilisers, irrigation, technology products, insurance, logistics and allied activities.
13. Care to conserve water resources: India has 4% of water resources of the world and 16% of population. A majority of farmers depend on rain water for irrigation. Rain water harvesting and the use of drip irrigation should be institutionalised to conserve water resources.
14. Care to create land funds and rental market: Small farm holders should merge their land holdings and create a land fund. It should be institutionalised in the form of mutual funds to give the benefit of professional management and technology driven products to small farmers.
15. Care to create farm to serve products: The cooperatives of small farmers should be provided institutional support to set up food processing units to convert their raw output into finished processed food ready to serve. Instead of raw output, branded finished processed products should leave the farm. The online market places and offline market places should be explored for the marketing of these farms to serve products.
16. Care to digitise land records: One of the biggest challenges in bringing efficiency in operations and adoption of technology is lack of tradability and traceability of land due to the outdated manner in which land records are maintained. The use of biometric identification tools should be enhanced and all land records should be digitised.

10.9 CONCLUSION

This chapter has contributed to identifying the potential of existing technological infrastructure and capability in India and the manner in which it can be put together to resolve issues of the agriculture sector. The chapter has provided the CARE model for integrating various elements of technological, operational and financial support systems to connect with the farmers in a more effective and efficient manner. The operational model of the caring approach, shared land, and technological resources will enhance the access to technology of small and marginal farmers. Technology has changed the way we live, entertain, commute, connect, safeguard ourselves, and do business. The agriculture sector has yet to fully reap the rewards of technology. Blockchain has the potential to transform the lives of farmers. Blockchain can bring cost effective and efficient credit delivery, land records upkeep, unlocking value in small farms by developing the land rental market, crop insurance, and better price recovery with a revamped supply chain, supported by blockchain. Blockchain removes the role of middlemen and enables peer to peer transactions. Existing digital infrastructure provides an opportunity for start-ups in blockchain to offer services by leveraging the existing digital foundation. The size of the population depending on agriculture provides scope to offer blockchain based services to a large population base and achieve economy of scale in operations, marketing, and financing. Small farmers should be mentored to diversify into high value crops and organic farming. An Agritech revolution in the form of Agriculture 5.0 is the current need to enhance the incomes of farmers and unlock the value of Indian agriculture. Blockchain has the potential to transform Indian farmers from poverty to prosperity and contribute to realising the goal of a $ 5 trillion economy with Agriculture 5.0. Instead of the failed subsidy model, policy makers should encourage creativity, efficiency by adopting CARE model to enhance adoption of technology in Indian agriculture, and make farmers a prosperous community.

DECLARATION OF CONFLICTING INTEREST

It is declared that authors have no potential conflicts of interest with respect to the research, authorship and/or publication of this article.

FUNDING

The authors received no financial support for the research, authorship and/or publication of this article.

REFERENCES

Agnext. 2020. “Demo-our technology,” agnext.com/demo/our-technology/ accessed August 26.

Agrichain. 2020. “Agrichain for growers,” platform.agrichian.com/ accessed October 7,

Anjum, B. and Tiwari, R. 2012a. “Role of information technology in women empowerment,” *Excel International Journal of Multidisciplinary Management Studies*, 2, 1: 226–233.

Anjum, B. and Tiwari, R. 2012b. "An exploratory study of supply side issues in Indian higher education," *Asia Pacific Journal of Marketing and Management Review,* 1, 1: 14–24.

Anjum, B. and Tiwari, R. 2012c. "Role of private sector banks for financial inclusion," *Zenith International Journal of Multidisciplinary Research,* 2, 1: 270–280.

Anjum, B. and Tiwari, R. 2012d. "Role of manufacturing industries in India for inclusive growth," *ZENITH International Journal of Business Economics & Management Research,* 2, 1: 97–104.

Apte, S. and Petrovsky, N. 2016. "Will blockchain technology revolutionize excipient supply chain management?" *Journal of Excipients and Food Chemical,* 3, 7: 76–78.

AUS. 2020. "About us," aus.co.in/about-us/ accessed September 8, 2020.

Bera, S. 2018. "Small and marginal farmers own just 47.3% of crop area, shows farm census," *Live Mint,* October 1, 2018. www.livemint.com/Politics/k90ox8AsPMdyPDuykv1eWL/Small-and-marginal-farmers-own-just-473-of-crop-area-show.html

Bright, J. 2019. "WorldCover raises $6M round for emerging markets climate insurance," Accessed October 14, 2020. techcrunch.com/2019/05/03/worldcover-raises-6m-round-for-emerging-markets-climate-insurance/.

CEMA. 2018. "Priorities." Accessed October 12, 2020. www.cema-agri.org/index.php?option=com_content&view=category&id=10&Itemid=102.

Chakravorti, B. and Chaturvedi R. S. 2019. "How effective is India's government, compared with those in other emerging markets?" *Harvard Business Review,* (2019), accessed August 3, 2020. hbr.org/2019/05/how-effective-is-indias-government-compared-with-those-in-other-emerging-markets

Chakravarty, S. 2018. "Reimagining Indian agriculture: How technology can change the game for Indian farmers?" *Business Word,* August 25, 2018, www.businessworld.in/article/Reimagining-Indian-Agriculture-How-technology-can-change-the-game-for-Indian-farmers-/24-11-2018-164502/

Chakravorty, S., Chandrasekhar, S. and Naraparaju, K. 2016. "Income generation and inequality in India's agricultural sector: The consequences of land fragmentation," *Indira Gandhi Institute of Development Research,* accessed September 9, 2020. www.igidr.ac.in/pdf/publication/WP-2016-028.pdf

Chaudhury, D. R. 2019. "PM Modi presents '5-I' vision to maximise tech for social benefits at G20," *The Economic Times,* June 29, 2019. economictimes.indiatimes.com/news/politics-and-nation/pm-modi-presents-5-i-vision-to-maximise-tech-for-social-benefits-at-g20/articleshow/69997726.cms?utm_source=contentofinterest&utm_medium=text&utm_campaign=cppst

Choudhuri, S., Dixit, R. and Tiwari, R. 2015. "Issues and challenges of Indian aviation industry: A case study," *International Journal of Logistics & Supply Chain Management Perspectives,* 4, 1: 1557–1562.

Christensen, C. M. and Raynor, M. E. 2003. The innovator's solution: Creating and sustaining successful growth. Harvard Business Press,

DCunha, S. D. 2018. "For India's farmers it's Agtech Startups, not government, that is key," *Forbes.* www.forbes.com/sites/suparnadutt/2018/01/08/for-indias-farmers-its-agtech-startups-not-government-that-is-key/#5da8fbe11c6e

Debroy, B. and Jain, S. 2017. "Strengthening arbitration and its enforcement in India resolve in India," Working Papers id:11752.eSocialSciences.

Demeter. 2020. "Reinventing agriculture through blockchain," accessed October 21, 2020. demeter.life/

Etherisc. 2020. "Etherisc," accessed October 19, 2020. etherisc.com/

Feder, G., Just, R., and Zilberman, D. 1985. "Adoption of agricultural innovations in developing countries: A survey," *Economic Development and Cultural Change,* 33, 2 : 255–298.

Etherisc 2020. "Make Insurance Fair and Accessible," accessed October 10, 2020. etherisc.com/

FAO 2020. "Resilience, Food and Agricultural Organization," United Nations, accessed October 8, 2020. www.fao.org/resilience/news-events/detail/en/c/395608/

Haq, Z. 2019. "Food India wastes can feed all of Bihar for a year, shows govt study," *Hindustan Times*. November 4, 2019. www.hindustantimes.com/india-news/food-india-wastes-can-feed-all-of-bihar-for-a-year-shows-govt-study/story-qwV3C9YnJAoXn83b3htmsK.html

Hertz, L. 2019. "How will blockchain agriculture revolutionize the food supply from farm to plate?" accessed July 13, 2020. hackernoon.com/how-will-blockchain-agriculture-revolutionize-the-food-supply-from-farm-to-plate-f8fe488d9bae.

Huang, J., Wang, X., and Qiu, H. 2012. "Small-scale farmers in China in the face of modernisation and globalisation, knowledge programme small producer agency in the globalised market," accessed May 2, 2020. pubs.iied.org/pdfs/16515IIED.pdf

HuffPost 2017. "80% of India's poor mainly depend on farming, how can they be taxed, asks NITI Ayog chairman," accessed September, 12 2020. www.huffingtonpost.in/2017/04/29/80-of-indias-poor-mainly-depend-on-farming-how-can-they-be-ta_a_22061287/

India Brand Equity Foundation (IBEF) 2020. "Agriculture in India: Information about Indian agriculture & its importance," accessed June 17, 2020. www.ibef.org/archives/detail/b3ZlcnZpZXcmMzcwOTUmODY=

ITCPortal 2020. "E-Choupal." Accessed October 19, 2020. www.itcportal.com/businesses/agri-business/e-choupal.aspx

Kapoor, M. and Usmani, A. 2019. "Startup street: Five Indian startups most likely to turn unicorns in 2019," accessed September 11, 2020. www.bloombergquint.com/technology/startup-street-five-indian-startups-most-likely-to-turn-unicorns-in-2019

Khan, S. and Kannapiran, T. 2019. "Indexing issues in spatial big data management," *In International Conference on Advances in Engineering Science Management & Technology (ICAESMT)*, 2019, Uttaranchal University, Dehradun, India.

Khan, S., Redha Qader, M., Thirunavukkarasu, K., and Abimannan S. 2020. "Analysis of business intelligence impact on organizational performance," *In 2020 International Conference on Data Analytics for Business and Industry: Way Towards a Sustainable Economy (ICDABI)*, pp. 1–4. IEEE, Sakheer, Bahrain.

Khan, S., Al-Dmour, A., Bali, V., Rabbani, M. R., and Thirunavukkarasu, K. 2021. "Cloud computing based futuristic educational model for virtual learning," *Journal of Statistics and Management Systems*, 24, 2: 357–385.

Khem, C., Tiwari, R., and Phuyal, M. 2017. "Economic growth and unemployment rate: An empirical study of Indian economy," *Pragati: Journal of Indian Economy*, 4, 2: 130–137, DOI: 10.17492/pragati.v4i02.11468.

Joblaew, P., Sirisunyaluck, R., Kanjina, S., Chalermphol, J., and Prom, C. 2020. "Factors affecting farmers' adoption of rice production technology from the collaborative farming project in Phrae province, Thailand," *International Journal of Agricultural Technology*. 15, 6: 901–912.

Kondoker, A. M. 2018. "Perception and adoption of a new agricultural technology: Evidence from a developing country," *Technology in Society*, 55: 126–135.

Leiker, A. 2018. "Make way for blockchain," Tromoxie.com. Accessed September 18, 2020. trimoxie.com/make-way-blockchain/

Maiervidorno. 2020. "Agriculture in India: The need for new technologies." Accessed October 16, 2020. www.maiervidorno.com/agriculture-india-need-new-technologies/

Mendonca, J. 2019. "India's agricultural farms get a technology lift," *The Economic Times*, July 26, 2019. tech.economictimes.indiatimes.com/news/internet/indias-agricultural-farms-get-a-technology-lift/70388635.

Mohapatra, T. 2016. "From the DG's Desk," ICAR Reporter, April-June 2016. www.icar.org.in/files/IR-April-June-2016.pdf

Mulraj, J. 2019. "Blockchain, not blockheads, can provide solutions to India's problems," *The Hindu Business Line*. December 6, 2019. www.thehindubusinessline.com/markets/blockchain-not-blockheads-can-provide-solutions-to-indias-problems/article30213014.ece#

Mukherjee, S. 2018. "Organic food exports surge but certification remains a major issue," *Business Standard*, March 28, 2018. www.business-standard.com/article/economy-policy/organic-food-exports-surge-certification-remains-a-major-issue-118032800261_1.html

Nakamoto, S. 2008. "Bitcoin: A peer-to-peer electronic cash system," accessed September 23 bitcoin.org/bitcoin.pdf

Narayanamoorthy, A. 2019. "Tap drip irrigation to save water," *The Hindu Business Line*. June 7, 2019. www.thehindubusinessline.com/opinion/tap-drip-irrigation-to-save-water/article27688289.ece

NITI Aayog. 2020. "BlockChain: The India strategy: Towards enabling ease of business, ease of living, and ease of commerce," Accessed October 20, 2020. //niti.gov.in/sites/default/files/2020-01/Blockchain_The_India_Strategy_Part_I.pdf

Padmanabhan, V. 2018 "The land challenge underlying India's farm crisis," *Live Mint*, October 15, 2018. www.livemint.com/Politics/SOG43o5ypqO13j0QflaawM/The-land-challenge-underlying-Indias-farm-crisis.html

Patel, D., Bothra, J., and Patel, V 2017. "Blockchain exhumed," *ISEA Asia Security and Privacy (ISEASP)*, 1–12. doi: 10.1109/ISEASP.2017.7976993.

Prakash, A. and Parija, P. 2019. "Here's why farmers matter so much to India's Modi," *Bloomberg*. Accessed July 8, 2020. www.bloomberg.com/news/articles/2019-01-31/why-election-goodies-await-india-s-struggling-farmers-quicktake

Press Information Bureau (pib). 2020. "Key highlights of economic survey 2019-20," Ministry of finance. Accessed October 23, 2020. pib.gov.in/newsite/PrintRelease.aspx?relid=197771

PWC. 2020. "Global FinTech Report 2017, redrawing the lines: FinTech's growing influence on financial services," accessed May 13, 2020. www.pwc.com/gx/en/industries/financial-services/assets/pwc-global-fintech-report-2017.pdf

Raju, B. Y. 2019. "Tenant farmers being left high and dry," *The Hindu Business Line*. January 24, 2019. www.thehindubusinessline.com/opinion/tenant-farmers-being-left-high-and-dry/article26081913.ece#.

Rana, A. and Tiwari, R. 2014. "MSME sector: Challenges and potential growth strategies," *International Journal of Entrepreneurship & Business Environment Perspectives*. 3, 4: 1428–1432.

Rani, A., Chaudhary, A., Sinha, N., Mohanty, M., and Chaudhary, R. 2019. "DRONE: The green technology for future agriculture," *Harit Dhara*, 2, 1: 3–6.

Rawat, A. 2020. "These 15 drone startups are flying high in India," accessed July 14, 2020. inc42.com/features/these-15-drone-startups-are-flying-high-in-indias-digital-sky/

Reddy, J. 2020. "Latest agriculture technologies in India, impact, advantages," accessed October 21, 2020. www.agrifarming.in/latest-agriculture-technologies-in-india-impact-advantages#:~:text=Importance%20of%20latest%20agricultural%20technologies,benefits%20of%20agricultural%20technology%20include%3B&text=Decreased%20use%20of%20water%20quantity,turn%20keeps%20food%20prices%20down

Reddy, N., Reddy, A., and Kumar, J. 2016. "A critical review on agricultural robots," *International Journal of Mechanical Engineering Technology*, 7, 4: 183–188.

Ripe. 2020. "About Us," accessed October 21, 2020. www.ripe.io/about

Rubio, V. and Mas, F. 2020. "From Smart Farming towards Agriculture 5.0: A Review on Crop Data Management," *Agronomy*. 10, 2: 207–227. doi:10.3390/agronomy10020207

Sally, M. 2018. "Global demand for Indian organic food products on constant increase," *The Economic Times*, October 23, 2018. economictimes.indiatimes.com/industry/cons-products/food/global-demand-for-indian-organic-food-products-on-constant-increase/articleshow/66330641.cms?utm_source=contentofinterest&utm_medium=text&utm_campaign=cppst

Sangal, P. P. 2018. "Farmers are poor due to low productivity of all major crops," *Financial Express*. February 1, 2018. www.financialexpress.com/opinion/farmers-are-poor-due-to-low-productivity-of-all-major-crops/1038918/

Seth, A. and Ganguly, K. 2020. "Digital technologies transforming Indian agriculture," accessed April 18, 2020. www.wipo.int/edocs/pubdocs/en/wipo_pub_gii_2017-chapter5.pdf

Sharma, H., Tiwari, R., and Anjum, B. 2016. "Issues and challenges of affiliation system in Indian higher education," *EXCEL International Journal of Multidisciplinary Management Studies*. 3, 12: 232–240.

Shrivastava, S. 2020. "Top 10 ingenious drone startups in India 2020" accessed October 19, 2020. www.analyticsinsight.net/top-10-ingenious-drone-startups-india-2020/

Sirisunyaluck, R., Singsin, P. and Kanjina, S. 2020. "Factors influencing the adoption of climate change adaptation samong rice growers in Doi Saket District, Chiang Mai Province, Thailand," *International Journal of Agricultural Technology*, 16, 1: 129–142.

Solomon, O., Xavier, G., Joel, J., Duncan, O., and Hans, D. 2019. "Farmers' adoption of agricultural innovations: A systematic review on willingness to pay studies," *Outlook on Agriculture*, 49, 3: 187–203, October 2019.

Startup Insight. 2020. "8 blockchain startups disrupting the agricultural industry," accessed September 2, 2020. www.startus-insights.com/innovators-guide/8-blockchain-startups-disrupting-the-agricultural-industry/

Telecom Regulatory Authority of India (TRAI) 2020. "Indian telecom services performance indicator report for the quarter ending July-September, 2019," accessed October 5, 2020. main.trai.gov.in/sites/default/files/PR_No.04of2020.pdf

Tiwari, R. and Anjum, B. 2014. "Role of higher education institutions and industry academia collaboration for skill enhancement," *Journal of Business Management & Social Sciences Research*. 3, 11: 27–34.

Tiwari, R., Khem Chand, and Anjum, B. 2020. "Crop insurance in India: A review of pradhan mantri fasal bima yojana (PMFBY)," *FIIB Business Review*, 9, 4: 249–255, doi.org/10.1177/2319714520966084

Tomu. 2020. "Blockchain for agriculture: How blockchain can revolutionize food supply from the farm to the plate," accessed June 24, 2020. medium.com/swlh/blockchain-for-agriculture-5b0a0baa0aa3

TataTrsuts. 2020. "mKrishi," accessed July 9, 2020. www.tatatrusts.org/our-work/livelihood/agriculture-practices/mkrishi

Traxcn. 2020. "Top Agricultural Drones Startups," accessed October 7, 2020. tracxn.com/d/trending-themes/Startups-in-Agriculture-Drones

Trringo 2020. "About Us," accessed September 18, 2020. www.trringo.com/about-us.php

Udas, R. 2020. "Indian Agriculture goes Hi-Tech with New Technologies like AI, ML and IoT," accessed October 10, 2020. www.expresscomputer.in/features/indian-agriculture-goes-hi-tech-with-new-technologies-like-ai-ml-and-iot/45432/

Welvaert, M. 2020. "Food waste: A global challenge, a local solution," Committee of World Food Security. Accessed September 8, 2020. www.fao.org/cfs/home/blog/blog-articles/article/en/c/449010/

World Bank. 2007. *India: Land Policies for Growth and Poverty Reduction*, New Delhi, Oxford University Press.

WorldCovr. 2020. "Crop Insurance that works," accessed August 12, 2020. www.worldcovr.com/.

Xiang, X., Li, Q., Khan, S., and Khalaf, O. I. 2021. "Urban water resource management for sustainable environment planning using artificial intelligence techniques," *Environmental Impact Assessment Review* 86: 106515.

Yigezu, A., Amin, M., Tamer, E., Aden, A., Atef, H., Yaseen, K., and Stephen, L. 2018. "Enhancing adoption of agricultural technologies requiring high initial investment among smallholders," *Technological Forecasting and Social Change*. 134: 199–206.

Index